P9-APV-852

One of the best ways to raise your Earth Science regents grade is to diligently do four previous exams with complete comprehension. When you are done, you will have a working understanding of hundreds of questions and the supporting concepts. Many of the questions, in one way or another, will appear in this year's regents exam. This is the purpose of this review book. But the trick is to do the exams in earnest; taking your time, checking over our brief but concise explanations until it makes sense, and revisiting the ones you answer incorrectly days later to check your understanding of the correct answer.

Timing is essential. Don't wait until the last week. We suggest that you start working on these regents exams early, doing 20 to 30 questions a day. Star the ones you need to revisit, underline important information, and have a good knowledge of what is in the **RT- Reference Tables**. We suggest that you use the RT found in the back of this booklet or one that your teacher might have provided for you. Many points can be gained by knowing where in the RT an answer is found.

So as the limestone said to the bedrock; don't take the regents for granite. Rather, work hard and your grade will improve.

Best of luck

PRACTICE TESTS
for
PHYSICAL SETTING
REGENTS
EARTH SCIENCE

Answers Written By:

William Docekal

Science Teacher – Retired

Published by

TOPICAL REVIEW BOOK COMPANY

P. O. Box 328

Onsted, MI 49265-0328

1-800-847-0854

PHYSICAL SETTING
REGENTS
EARTH SCIENCE

Published by
TOPICAL REVIEW BOOK COMPANY
P. O. Box 328
Onsted, MI 49265-0328
1-800-847-0854

Answer all questions in this part.

Directions (1–35): For *each* statement or question, write in the space provided the *number* of the word or expression that, of those given, best completes the statement or answers the question. Some questions may require the use of the *Earth Science Reference Tables.*

1. Which list of three planets and Earth's Moon is arranged in order of increasing equatorial diameter?
(1) Earth's Moon, Pluto, Mars, Mercury
(2) Pluto, Earth's Moon, Mercury, Mars
(3) Mercury, Mars, Earth's Moon, Pluto
(4) Mars, Mercury, Pluto, Earth's Moon   1 _____

2. If Earth's axis were tilted 35° instead of 23.5°, the average temperatures in New York State would most likely
(1) decrease in both summer and winter
(2) decrease in summer and increase in winter
(3) increase in summer and decrease in winter
(4) increase in both summer and winter   2 _____

3. Which star has a higher luminosity and a lower temperature than the Sun?
(1) *Rigel*   (2) *Barnard's Star*   (3) *Alpha Centauri*   (4) *Aldebaran*   3 _____

4. Starlight from distant galaxies provides evidence that the universe is expanding because this starlight shows a shift in wavelength toward the
(1) red-light end of the visible spectrum
(2) blue-light end of the visible spectrum
(3) ultraviolet-ray end of the electromagnetic spectrum
(4) gamma-ray end of the electromagnetic spectrum   4 _____

5. On which day of the year would the intensity of insolation at Kingston, New York, most likely be greatest?
(1) March 21   (2) June 21   (3) September 23   (4) December 21   5 _____

6. The accompanying diagram represents the elliptical orbit of a moon revolving around a planet. The foci of this orbit are the points labeled F_1 and F_2. What is the approximate eccentricity of this elliptical orbit?
(1) 0.3   (3) 0.7
(2) 0.5   (4) 1.4

$l = 7.5$ cm
$d = 5.5$ cm

Moon

Planet
F_1 F_2

Drawing not to scale

6 _____

7. The coldest climates on Earth are located at or near the poles primarily because Earth's polar regions
(1) receive mostly low-angle insolation
(2) receive less total yearly hours of daylight
(3) absorb the greatest amount of insolation
(4) are usually farthest from the Sun   7_____

8. Compared to an inland location, a location on an ocean shore at the same elevation and latitude is likely to have
(1) cooler winters and cooler summers
(2) cooler winters and warmer summers
(3) warmer winters and cooler summers
(4) warmer winters and warmer summers   8_____

9. The accompanying diagram represents a Foucault pendulum swinging freely for 8 hours. The Foucault pendulum appears to gradually change its direction of swing due to Earth's
(1) orbit around the Sun
(2) curved surface
(3) tilted axis
(4) spin on its axis

9_____

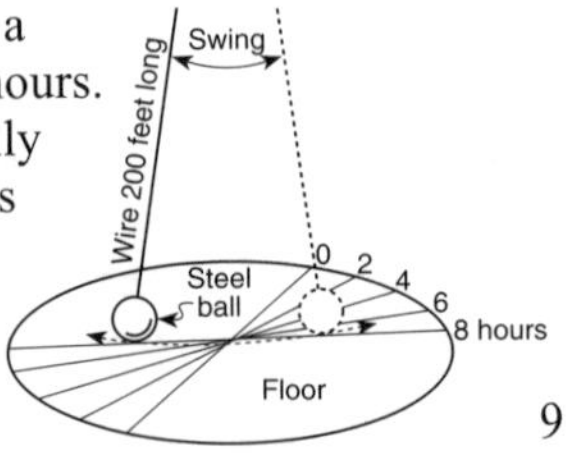

10. The accompanying diagram shows the noontime shadows cast by a student and a tree. If the time is solar noon and the student is located in New York State, in what direction is the student facing?
(1) north   (3) east
(2) south   (4) west

10_____

11. The accompanying diagram represents the Moon in its orbit, as viewed from above Earth's North Pole. Position 1 represents a specific location of the Moon in its orbit. Which phase of the Moon will be seen from Earth when the Moon is at position 1?

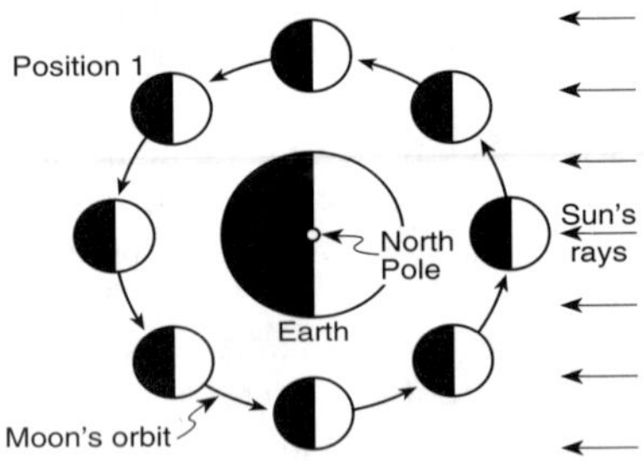

(Not drawn to scale)

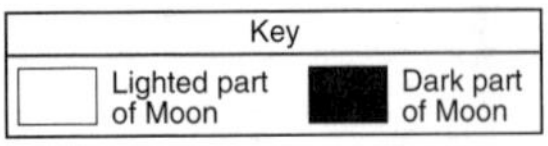

Key	
Lighted part of Moon	Dark part of Moon

(1)   (2)   (3)   (4)   11_____

12. During nighttime cooling, most of the energy radiated by Earth's oceans into space is
(1) ultraviolet rays
(2) gamma rays
(3) visible light rays
(4) infrared rays

12 _____

13. Landscapes with horizontal bedrock structure, steep slopes, and high elevations are classified as
(1) plateau regions
(2) plain regions
(3) lowland regions
(4) mountain regions

13 _____

14. Which map view best represents the pattern of isobar values, in millibars, and the pattern of wind flow, shown by arrows, at Earth's surface surrounding a Northern Hemisphere low-pressure center?

(1)　　　　(2)　　　　(3)　　　　(4)

14 _____

15. An observer measured the air temperature and the dewpoint and found the difference between them to be 12°C. One hour later, the difference between the air temperature and the dewpoint was found to be 4°C. Which statement best describes the changes that were occurring?
(1) The relative humidity was decreasing and the chance of precipitation was decreasing.
(2) The relative humidity was decreasing and the chance of precipitation was increasing.
(3) The relative humidity was increasing and the chance of precipitation was decreasing.
(4) The relative humidity was increasing and the chance of precipitation was increasing.

15 _____

16. Which two ocean currents are both warm currents that primarily flow away from the equator?
(1) Guinea Current and Labrador Current
(2) Brazil Current and Agulhas Current
(3) Alaska Current and Falkland Current
(4) Canaries Current and Gulf Stream Current

16 _____

17. Which surface soil conditions allow the most infiltration of rainwater?
(1) steep slope and permeable soil
(2) steep slope and impermeable soil
(3) gentle slope and permeable soil
(4) gentle slope and impermeable soil

17 _____

18. The accompanying map shows a river emptying into an ocean, producing a delta. Which graph best represents the relationship between the distance from the river delta into the ocean and the average size of sediments deposited on the ocean floor?

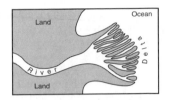

(1)

(2)

(3)

(4)

18 _____

19. Which process could lead directly to the formation of pumice rock?
(1) precipitation of minerals from evaporating seawater
(2) metamorphism of unmelted rock material
(3) deposition of quartz sand
(4) explosive eruption of lava from a volcano

19 _____

20. The accompanying rock shown has a foliated texture and contains the minerals amphibole, quartz, and feldspar arranged in coarse-grained bands. Which rock is shown?
(1) slate   (2) dunite   (3) gneiss   (4) quartzite

20 _____

21. The accompanying photograph shows the intergrown crystals of a pegmatite rock. Which characteristic provides the best evidence that this pegmatite solidified deep underground?
(1) low density   (3) felsic composition
(2) light color   (4) very coarse texture

(Actual size)

21 _____

22. Which igneous rock, when weathered, could produce sediment composed of the minerals potassium feldspar, quartz, and amphibole?
(1) gabbro   (2) granite   (3) andesite   (4) basalt

22 _____

Base your answers to questions 23 and 24 on the accompanying photograph. The photograph shows several broken samples of the same colorless mineral.

23. Which physical property of this mineral is most easily seen in the photograph?
(1) fracture   (2) hardness   (3) streak   (4) cleavage

23 _____

24. Which mineral is most likely shown in the photograph?
(1) quartz   (2) calcite   (3) galena   (4) halite

24 _____

25. The large coal fields found in Pennsylvania provide evidence that the climate of the northeastern United States was much warmer during the Carboniferous Period. This change in climate over time is best explained by the
(1) movements of tectonic plates
(2) effects of seasons
(3) changes in the environment caused by humans
(4) evolution of life   25 _____

26. What is the inferred temperature at the boundary between Earth's stiffer mantle and outer core?
(1) 2,500°C   (2) 4,500°C   (3) 5,000°C   (4) 6,200°C   26 _____

27. Which color of the visible spectrum has the *shortest* wavelength?
(1) violet   (2) blue   (3) yellow   (4) red   27 _____

28. The accompanying diagram shows the interaction of two tectonic plates. The type of plate boundary represented in the diagram most likely exists between the
(1) Antarctic Plate and the African Plate
(2) South American Plate and the Nazca Plate
(3) South American Plate and the African Plate
(4) Antarctic Plate and the Indian-Australian Plate   28 _____

(Not drawn to scale)

29. When 1 gram of liquid water at 0° Celsius freezes to form ice, how many total calories of heat are lost by the water?
(1) 1   (2) 0.5   (3) 80   (4) 540   29 _____

30. The shore of which New York State body of water has large amounts of metamorphic bedrock exposed at the surface?
(1) western shore of Lake Champlain
(2) eastern shore of Lake Erie
(3) southern shore of Long Island Sound
(4) southern shore of Lake Ontario   30 _____

31. Which graph best shows the radioactive decay of carbon-14?

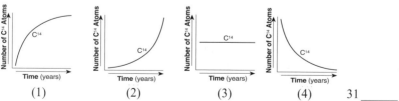

(1)   (2)   (3)   (4)   31 _____

32. Approximately what percentage of the estimated age of Earth does the Cenozoic Era represent?
(1) 1.4%   (2) 5.0%   (3) 11.9%   (4) 65.0%   32 _____

33. The accompanying geologic cross section shows the geologic age of two rock layers separated by an unconformity. The unconformity at the bottom of the Silurian rock layer indicates a gap in the geologic time record. What is the *minimum* time, in millions of years, shown by the gap?

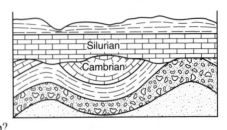

(1) 13   (2) 47   (3) 101   (4) 126   33 _____

34. The accompanying diagram represents a map view of a stream drainage pattern.

Which underlying bedrock structure most likely produced this stream drainage pattern?

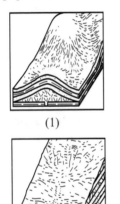

(1)   (3)

(2)   (4)   34 _____

35. Compared to a maritime tropical air mass, a maritime polar air mass has a
(1) higher temperature and more water vapor
(2) higher temperature and less water vapor
(3) lower temperature and more water vapor
(4) lower temperature and less water vapor   35 _____

Part B–1

Answer all questions in this part.

Directions (36–50): For *each* statement or question, write in the space provided the *number* of the word or expression that, of those given, best completes the statement or answers the question. Some questions may require the use of the *Earth Science Reference Tables.*

Base your answers to questions 36 through 38 on the accompanying map, which shows sea-level air pressure, in millibars, for a portion of the eastern coast of North America. Points *A, B, C,* and *D* are sea-level locations on Earth's surface.

Sea-Level Air Pressures

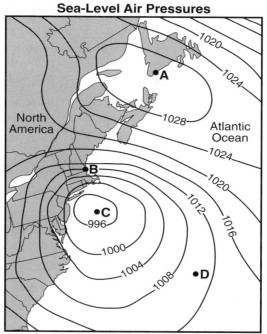

36. Which weather instrument was used to measure the air pressures?
(1) thermometer   (3) sling psychrometer
(2) wind vane   (4) barometer   36 _____

37. Which location most likely recorded the highest wind speed?
(1) *A*   (2) *B*   (3) *C*   (4) *D*   37 _____

38. The air pressure recorded at point *D* was most likely
(1) 1014 mb   (2) 1012 mb   (3) 1010 mb   (4) 1006 mb   38 _____

Base your answers to questions 39 through 41 on the newspaper article shown below and on your knowledge of Earth science.

Fossilized Jellyfish Found in Wisconsin

Fossil hunters have unearthed the largest collection of fossilized jellyfish ever discovered, including the largest fossilized jellyfish ever found.

The remains of soft-bodied animals such as jellyfish are relatively rare because they don't have bones, fossil dealer Dan Damrow, James W. Hagadorn of the California Institute of Technology and Robert H. Dott Jr. of the University of Wisconsin at Madison noted in describing the find in the journal Geology.

About a half-billion years ago, during the Cambrian period, the quarry in Mosinee, Wis., where the deposits were found was a small lagoon. The jellyfish apparently died when they were washed up by a freak tide or storm, the researchers said. The jellyfish remains were probably preserved because of a lack of erosion from sea water and wind, and a lack of scavengers, the researchers concluded.

"It is very rare to discover a deposit which contains an entire stranding event of jellyfish," Hagadorn said. "These jellyfish are not just large for the Cambrian, but are the largest jellyfish in the entire fossil record."

Washington Post, January 2002

39. These fossilized jellyfish were most likely discovered in which type of rock?
(1) sandstone   (2) granite   (3) pumice   (4) slate   39_____

40. Which two marine organisms most likely lived at the same time as these jellyfish?
(1) crinoids and dinosaurs   (3) brachiopods and gastropods
(2) ammonoids and placoderm fish (4) amphibians and eurypterids 40_____

41. Which evidence would lead scientists to suspect that a tide or storm had washed up these jellyfish on a beach?
(1) Primitive life existed on land 500 million years ago.
(2) The rock containing the jellyfish fossils has distorted crystal structure.
(3) Treeroot fossils appear to have been pitted and folded.
(4) Large ripple marks were found in the fossil-containing rock layers.   41_____

Base your answers to questions 42 through 44 on the topographic map below. Points *A, B, C, D,* and *X* represent locations on the map. Elevations are measured in feet.

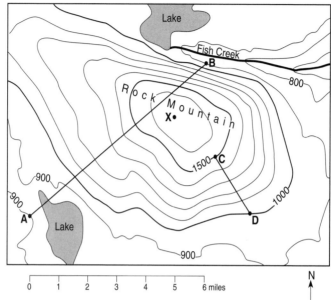

42. What is the highest possible elevation of point **X** on Rock Mountain?
(1) 1,599 ft   (2) 1,600 ft   (3) 1,601 ft   (4) 1,699 ft   42 _____

43. What is the average gradient of the slope along straight line *CD*?
(1) 100 ft/mi   (2) 250 ft/mi   (3) 500 ft/mi   (4) 1,000 ft/mi   43 _____

44. Which cross section best represents the profile along straight line *AB*?

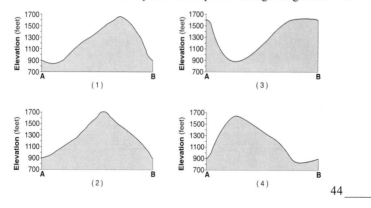

44 _____

Base your answers to questions 45 through 47 on the map below, which shows watershed regions of New York State.

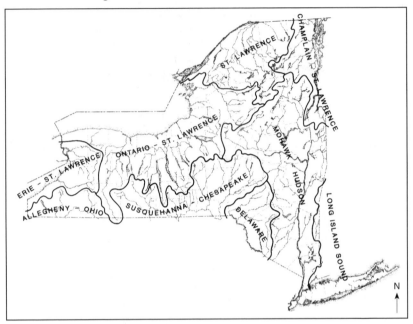

45. On which type of landscape region are both the Susquehanna-Chesapeake and the Delaware watersheds located?
(1) plain   (2) plateau   (3) mountain   (4) lowland   45 _____

46. In which watershed is the Genesee River located?
(1) Ontario-St. Lawrence   (3) Mohawk-Hudson
(2) Susquehanna-Chesapeake   (4) Delaware   46 _____

47. Most of the surface bedrock of the Ontario-St. Lawrence watershed was formed during which geologic time periods?
(1) Precambrian and Cambrian
(2) Ordovician, Silurian, and Devonian
(3) Mississippian, Pennsylvanian, and Permian
(4) Triassic, Jurassic, and Cretaceous   47 _____

Base your answers to questions 48 through 50 on the accompanying diagram, which represents an exaggerated view of Earth revolving around the Sun. Letters *A, B, C,* and *D* represent Earth's location in its orbit on the first day of each of the four seasons.

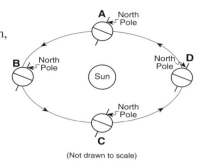

(Not drawn to scale)

48. Which location in Earth's orbit represents the first day of fall (autumn) for an observer in New York State?
(1) *A*   (2) *B*   (3) *C*   (4) *D*   48 _____

49. Earth's rate of revolution around the Sun is approximately
(1) 1° per day
(2) 360° per day
(3) 15° per hour
(4) 23.5° per hour   49 _____

50. Which observation provides the best evidence that Earth revolves around the Sun?
(1) Stars seen from Earth appear to circle *Polaris*.
(2) Earth's planetary winds are deflected by the Coriolis effect.
(3) The change from high ocean tide to low ocean tide is a repeating pattern.
(4) Different star constellations are seen from Earth at different times of the year.   50 _____

Part B–2

Answer all questions in this part.

Directions (51–65): **Record your answers in the spaces provided. Some questions may require the use of the *Earth Science Reference Tables*.**

Base your answers to questions 51 through 53 on the cross section below, which shows a typical cold front moving over New York State in early summer.

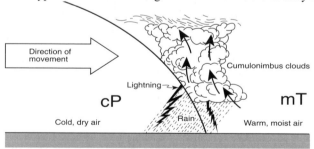

51. Explain why the warm, moist air is rising at the frontal boundary. [1]

_____

_____

52. State *one* process that causes clouds to form in this rising air. [1]

_____

53. Central Canada was the geographic source region for the cP air mass shown in the cross section. Identify the most likely geographic source region for the mT air mass shown in the cross section. [1]

_____

Base your answers to questions 54 through 57 on the accompanying map and on your knowledge of Earth science. The map shows the location of the epicenter,⊗, of an earthquake that occurred on April 20, 2002, about 29 kilometers southwest of Plattsburgh, New York.

54. State the latitude and longitude of this earthquake epicenter. Express your answers to the *nearest tenth of a degree* and include the compass directions. [1]

Latitude: _____

Longitude: _____

55. What is the *minimum* number of seismographic stations needed to locate the epicenter of an earthquake? [1]

_____

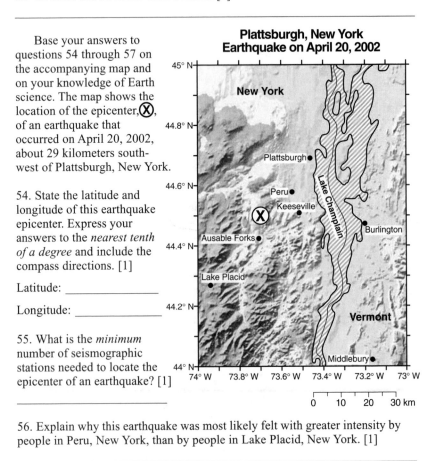

**Plattsburgh, New York
Earthquake on April 20, 2002**

56. Explain why this earthquake was most likely felt with greater intensity by people in Peru, New York, than by people in Lake Placid, New York. [1]

_____

_____

57. A seismic station located 1,800 kilometers from the epicenter recorded the *P*-wave and *S*-wave arrival times for this earthquake. What was the difference in the arrival time of the first *P*-wave and the first *S*-wave? [1]

_____ min _____ sec

Base your answers to questions 58 through 60 on the cross section of the accompanying bedrock outcrop and on your knowledge of Earth science. Index fossils found in some of the rock units are shown. The rock units are labeled I through IX.

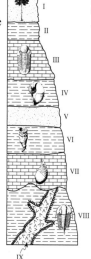

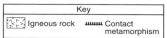

Key	
Igneous rock	Contact metamorphism

58. In the space below, number the relative age of rock units VII, VIII, and IX from 1 to 3, with number 1 indicating the oldest rock and number 3 indicating the youngest rock. [1]

Rock unit VII: _____

Rock unit VIII: _____

Rock unit IX: _____

59. The fossil shown in rock unit VIII is a member of an extinct group of fossils. State *two* other index fossils that are also members of the same group of extinct fossils. [1]

_____ and _____

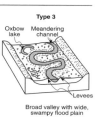

60. Based on the fossils shown in the limestone and shale layers, state the type of environment in which these sedimentary rocks were deposited. [1]

Base your answers to questions 61 through 63 on the accompanying block diagrams, which show three types of streams with equal volumes.

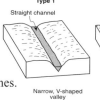
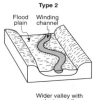

61. Explain how the differences between the type 1 and type 3 stream channels indicate that the average velocities of the streams are different. [1]

_____

_____

62. Explain why the outside of the curve of a meandering channel experiences more erosion than the inside of the curve. [1]

_____

_____

63. Explain how the cobbles and pebbles that were transported by these streams became smooth and rounded in shape. [1]

_____

_____

Base your answers to questions 64 and 65 on the accompanying station model, which shows the weather conditions at Rochester, New York, at 4 p.m. on a particular day in June.

64. What was the actual barometric pressure, according to the station model, to the *nearest tenth of a millibar*? [1] _____ mb

65. The winds shown by this station model were blowing from which compass direction and at what wind speed? [1]

From the _____ at _____ knots

Part C

Answer all questions in this part.

Directions (66–83): Record your answers in the spaces provided. Some questions may require the use of the *Earth Science Reference Tables.*

Base your answers to questions 66 and 67 on the accompanying data table, which lists the apparent diameter of the Sun, measured in minutes and seconds of a degree, as it appears to an observer in New York State. (Apparent diameter is how large an object appears to an observer.)

66. On the grid provided, graph the data shown on the table by marking with a dot the apparent diameter of the Sun for *each* date listed and connecting the dots with a smooth, curved line. [2]

Apparent Diameter of the Sun During the Year

Date	Apparent Diameter (' = minutes " = seconds)
January 1	32'32"
February 10	32'25"
March 20	32'07"
April 20	31'50"
May 30	31'33"
June 30	31'28"
August 10	31'34"
September 20	31'51"
November 10	32'18"
December 30	32'32"

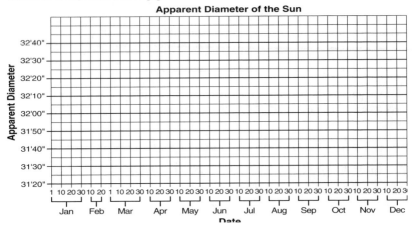

Apparent Diameter of the Sun

67. Explain why the apparent diameter of the Sun changes throughout the year as Earth revolves around the Sun. [1]

_____

_____

_____

Base your answers to questions 68 through 71 on the passage below and on your knowledge of Earth science.

Watching the Glaciers Go

Mountain glaciers and ice caps in tropical areas of the world are melting fast and may vanish altogether by the year 2020. That was the chilling news last year from Lonnie Thompson, a geologist at Ohio State University's Byrd Polar Research Center who has been studying icy areas near the equator in South America, Africa, and the Himalayas for two decades.

It doesn't take a glacier scientist to see the changes. In 1977, when Thompson visited the Quelccaya ice cap in Peru, it was impossible not to notice a schoolbus-size boulder stuck in its grip. When Thompson returned in 2000, the rock was still there but the ice wasn't — it had retreated far into the distance.

Most scientists believe the glaciers are melting because of global warming — the gradual temperature increase that has been observed with increasing urgency during the past decade. Last year a panel of the nation's top scientists, the National Research Council, set aside any lingering skepticism about the phenomenon, concluding definitively that average global surface temperatures are rising and will continue to do so.

"Watching the Glaciers Go," *Popular Science,* vol. #7, January 2002

68. State *one* greenhouse gas that is an excellent absorber of infrared radiation and may be responsible for global warming. [1]_____

69. Describe the arrangement of sediment deposited directly from glaciers. [1]

_____

70. Some glaciers currently exist near Earth's equator due to the cold, snowy climate of certain locations. Which type of landform exists where these glaciers occur? [1]

_____

71. Describe *one* action humans could take to reduce the global warming that is melting the Quelccaya ice cap. [1]

_____

_____

Base your answers to questions 72 through 75 on the map and cross section of the Finger Lakes Region shown below and on your knowledge of Earth science.

Finger Lakes Region of New York State

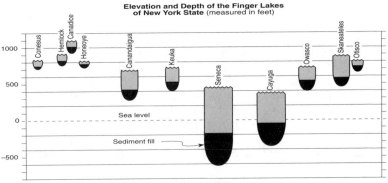

Elevation and Depth of the Finger Lakes of New York State (measured in feet)

72. According to the cross section, how thick from top to bottom is the sediment fill in Seneca Lake? [1] _____ ft

73. State *one* possible explanation for the north-south orientation of the Finger Lakes. [1]

_____

_____

74. During some winters, a few of the Finger Lakes remain unfrozen even though the land around the lakes is frozen. Explain how the specific heat of water can cause these lakes to remain unfrozen. [1]

_____

_____

75. Identify *two* processes that normally occur to form the type of surface bedrock found in the Finger Lakes Region. [1]

_____ and _____

Base your answers to questions 76 through 79 on the diagram below, which shows observations made by a sailor who left his ship and landed on a small deserted island on June 21. The diagram represents the apparent path of the Sun and the position of *Polaris,* as observed by the sailor on this island.

Sailor's Observations on the Deserted Island

76. On the diagram above, draw an arrow on the June 21 path of the Sun to show the Sun's direction of apparent movement from sunrise to sunset. [1]

77. The sailor was still on the island on September 23. On the diagram above, draw the Sun's apparent path for September 23, as it would have appeared to the sailor. Be sure your September 23 path indicates the correct altitude of the noon Sun and begins and ends at the correct points on the horizon. [2]

78. Based on the sailor's observations, what is the latitude of this island? Include the units and the compass direction in your answer. [1]   Latitude: _____

79. The sailor observed a 1-hour difference between solar noon on the island and solar noon at his last measured longitude onboard his ship. How many degrees of longitude is the island from the sailor's last measured longitude onboard his ship? [1]   _____

Base your answers to questions 80 through 83 on the world map shown below and on your knowledge of Earth science. Letters A through H represent locations on Earth's surface.

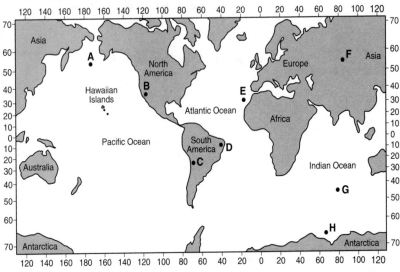

80. Explain why most earthquakes that occur in the crust beneath location B are shallower than most earthquakes that occur in the crust beneath location C. [1]

_____

_____

_____

81. Explain why location A has a greater probability of experiencing a major earthquake than location D. [1]

_____

_____

_____

82. Explain why a volcanic eruption is more likely to occur at location E than at location F. [1] _____

_____

_____

83. Explain why the geologic age of the oceanic bedrock increases from location G to location H.[1] _____

_____

_____

Part A

Answer all questions in this part.

Directions (1–35): For *each* statement or question, write in the space provided the *number* of the word or expression that, of those given, best completes the statement or answers the question. Some questions may require the use of the *Earth Science Reference Tables.*

1. Which object is closest to Earth?
(1) the Sun   (2) Venus   (3) the Moon   (4) Mars   1 _____

2. The accompanying diagram shows an observer on Earth measuring the altitude of *Polaris*. What is the latitude of this observer?
(1) 90° N
(2) 66.5° N
(3) 43° N
(4) 23.5° N

Zenith   Polaris
23.5°
66.5°
Horizon

2 _____

3. What is the minimum water velocity needed in a stream to maintain the transportation of the smallest boulder?
(1) 100 cm/sec   (2) 200 cm/sec   (3) 300 cm/sec   (4) 500 cm/sec   3 _____

4. Earth's early atmosphere formed during the Early Archean Era. Which gas was generally absent from the atmosphere at that time?
(1) water vapor   (2) carbon dioxide   (3) nitrogen   (4) oxygen   4 _____

5. The accompanying diagram shows the Moon orbiting Earth, as viewed from space above Earth's North Pole. The Moon is shown at eight positions in its orbit. Spring ocean tides occur when the difference in height between high tide and low tide is greatest. At which two positions of the Moon will spring tides occur on Earth?
(1) 1 and 5   (3) 3 and 7
(2) 2 and 6   (4) 4 and 8

Moon's orbit
Earth
North Pole
Sun's rays
(Not drawn to scale)

5 _____

6. Compared to other groups of stars, the group that has relatively low luminosities and relatively low temperatures is the
(1) Red Dwarfs   (3) Red Giants
(2) White Dwarfs   (4) Blue Supergiants   6 _____

7. Which sequence correctly lists the relative sizes from smallest to largest?
(1) our solar system, universe, Milky Way Galaxy
(2) our solar system, Milky Way Galaxy, universe
(3) Milky Way Galaxy, our solar system, universe
(4) Milky Way Galaxy, universe, our solar system

7 _____

8. The accompanying diagram represents a swinging Foucault pendulum. This pendulum will show an apparent change in the direction of its swing due to Earth's
(1) curved surface   (3) rotation
(2) tilted axis   (4) revolution

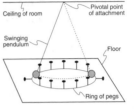

8 _____

9. The accompanying diagram shows the altitude of the Sun at solar noon on March 21, as seen by an observer at 42° N latitude. Compared to the altitude of the Sun observed at solar noon on March 21, the altitude of the Sun observed at solar noon on June 21 will be
(1) 15° higher in the sky
(2) 23.5° higher in the sky
(3) 42° higher in the sky
(4) 48° higher in the sky

9 _____

10. The accompanying diagram shows Earth's orbit around the Sun and different positions of the Moon as it travels around Earth. Letters A through D represent four different positions of the Moon. An eclipse of the Moon is most likely to occur when the Moon is at position
(1) A   (2) B   (3) C   (4) D

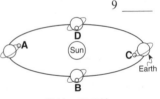

(Not drawn to scale)

10 _____

11. In the Northern Hemisphere, planetary winds blowing from north to south are deflected, or curved, toward the west. This deflection is caused by the
(1) unequal heating of land and water surfaces
(2) movement of low-pressure weather systems
(3) orbiting of Earth around the Sun
(4) spinning of Earth on its axis

11 _____

12. The accompanying table shows air-pressure readings taken at two cities, in the same region of the United States, at noon on four different days. The wind speed in the region between cities A and B was probably the greatest at noon on day
(1) 1   (3) 3
(2) 2   (4) 4

Air-Pressure Readings

Day	City A Air Pressure (mb)	City B Air Pressure (mb)
1	1004.0	1004.0
2	1000.1	1002.9
3	1000.2	1011.1
4	1010.4	1012.3

12 _____

13. If the base of a cloud is located at an altitude of 2 kilometers and the top of the cloud is located at an altitude of 8 kilometers, this cloud is located in the
(1) troposphere, only
(2) stratosphere, only
(3) troposphere and stratosphere
(4) stratosphere and mesosphere
13 ____

14. On a day with no wind, the air temperature outside a house is 10°C. The air temperature inside the house is 18°C. Which diagram best represents the air circulation pattern that is most likely to occur when a window of the house is first opened?

(1)   (2)   (3)   (4)   14 ____

15. Most of the Gulf Stream Ocean Current is
(1) warm water that flows southwestward
(2) warm water that flows northeastward
(3) cool water that flows southwestward
(4) cool water that flows northeastward
15 ____

16. Which event is the best example of erosion?
(1) breaking apart of shale as a result of water freezing in a crack
(2) dissolving of rock particles on a limestone gravestone by acid rain
(3) rolling of a pebble along the bottom of a stream
(4) crumbling of bedrock in one area to form soil
16 ____

17. Which graph best shows the relationship between the concentration of carbon dioxide in Earth's atmosphere and the amount of infrared radiation absorbed by the atmosphere?

(1)   (2)   (3)   (4)   17 ____

18. The accompanying diagram represents the direction of Earth's rotation as it appears from above the North Pole. Point X is a location on Earth's surface. The time at point X is closest to
(1) 6 a.m.   (2) 12 noon   (3) 6 p.m.   (4) 12 midnight

18 ____

19. Snowfall is rare at the South Pole because the air over the South Pole is usually
(1) rising and moist
(2) rising and dry
(3) sinking and moist
(4) sinking and dry
19 ____

20. The four streams shown on the topographic maps below have the same volume between *X* and *Y*. The distance from *X* to *Y* is also the same. All the maps are drawn to the same scale and have the same contour interval. Which map shows the stream with the greatest velocity between points *X* and *Y*?

| (1) | (2) | (3) | (4) | 20 ____ |

21. A student obtains a cup of quartz sand from a beach. A saltwater solution is poured into the sand and allowed to evaporate. The mineral residue from the saltwater solution cements the sand grains together, forming a material that is most similar in origin to
(1) an extrusive igneous rock   (3) a clastic sedimentary rock
(2) an intrusive igneous rock   (4) a foliated metamorphic rock   21 ____

22. Which coastal area is most likely to experience a severe earthquake?
(1) east coast of North America   (3) west coast of Africa
(2) east coast of Australia   (4) west coast of South America   22 ____

23. Which characteristic is most useful in correlating Devonian-age sedimentary bedrock in New York State with Devonian-age sedimentary bedrock in other parts of the world?
(1) color   (2) index fossils   (3) rock types   (4) particle size   23 ____

24. A seismic station 4000 kilometers from the epicenter of an earthquake records the arrival time of the first *P*-wave at 10:00:00. At what time did the first *S*-wave arrive at this station?
(1) 9:55:00   (2) 10:05:40   (3) 10:07:05   (4) 10:12:40   24 ____

25. Which statement correctly describes the density of Earth's mantle compared to the density of Earth's core and crust?
(1) The mantle is less dense than the core but more dense than the crust.
(2) The mantle is less dense than both the core and the crust.
(3) The mantle is more dense than the core but less dense than the crust.
(4) The mantle is more dense than both the core and the crust.   25 ____

26. Convection currents in the plastic mantle are believed to cause divergence of lithospheric plates at the
(1) Peru-Chile Trench   (3) Canary Islands Hot Spot
(2) Mariana Trench   (4) Iceland Hot Spot   26 ____

27. According to fossil evidence, which sequence shows the order in which these four life-forms first appeared on Earth?
(1) reptiles → amphibians → insects → fish
(2) insects → fish → reptiles → amphibians
(3) amphibians → reptiles → fish → insects
(4) fish → insects → amphibians → reptiles

27 ____

28. The accompanying fossil was found in surface bedrock in the eastern United States. Which statement best describes the formation of the rock containing this fossil?
(1) The rock was formed by the metamorphism of sedimentary rock deposited in a terrestrial environment during the Cretaceous Period.

(2) The rock was formed by the compaction and cementation of sediments deposited in a terrestrial environment during the Triassic Period.
(3) The rock was formed by the compaction and cementation of sediments deposited in a marine environment during the Cambrian Period.
(4) The rock was formed from the solidification of magma in a marine environment during the Triassic Period.

28 ____

29. The accompanying diagram shows an index fossil found in surface bedrock in some parts of New York State. In which New York State landscape region is this gastropod fossil most likely found in the surface bedrock?

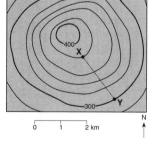

Maclurites

(1) Tug Hill Plateau   (3) Adirondack Mountains
(2) Allegheny Plateau   (4) Newark Lowlands

29 ____

30. The accompanying topographic map shows a hill. Points X and Y represent locations on the hill's surface. Elevations are shown in meters. What is the gradient between points X and Y?
(1) 40 m/km   (3) 100 m/km
(2) 80 m/km   (4) 120 m/km

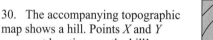

0   1   2 km

N

30 ____

31. The diagram shows a sling psychrometer.

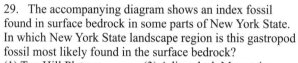

Wet bulb

Dry bulb

Based on the dry-bulb temperature and the wet-bulb temperature, what is the relative humidity?
(1) 66%   (2) 58%   (3) 51%   (4) 12%

31 ____

32. Arrows in the accompanying block diagram show the relative movement along a tectonic plate boundary. Between which two tectonic plates does this type of plate boundary exist?
(1) Nazca Plate and South American Plate
(2) Eurasian Plate and Indian-Australian Plate
(3) North American Plate and Eurasian Plate
(4) Pacific Plate and North American Plate

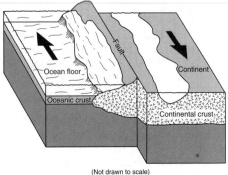

(Not drawn to scale)

32 _____

33. Which map shows the two correctly labeled air masses that frequently converge in the central plains to cause tornadoes?

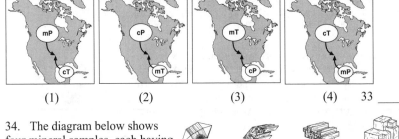

(1)   (2)   (3)   (4)   33 _____

34. The diagram below shows four mineral samples, each having approximately the same mass.

 Quartz   Amphibole   Pyroxene   Galena

If all four samples are placed together in a closed, dry container and shaken vigorously for 10 minutes, which mineral sample would experience the most abrasion?
(1) quartz   (2) amphibole   (3) pyroxene   (4) galena   34 _____

35. Which block diagram best represents a portion of a plateau?

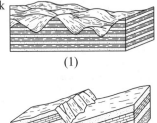

(1)   (3)

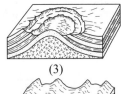

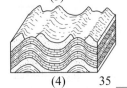

(2)   (4)   35 _____

Part B–1
Answer all questions in this part.

Directions (36–50): For *each* statement or question, write in the space provided the *number* of the word or expression that, of those given, best completes the statement or answers the question. Some questions may require the use of the *Earth Science Reference Tables*.

Base your answers to questions 36 through 38 on the accompanying graph, which shows the crustal temperature and pressure conditions under which three different minerals with the same chemical composition (Al_2SiO_5) crystallize.

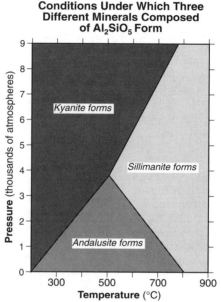

36. Under which crustal temperature and pressure conditions will andalusite form?
(1) 300°C and 6000 atmospheres   (3) 600°C and 4000 atmospheres
(2) 500°C and 2000 atmospheres   (4) 700°C and 8000 atmospheres   36 ____

37. Which mineral has a chemical composition most similar to andalusite, sillimanite, and kyanite?
(1) pyrite   (2) gypsum   (3) dolomite   (4) potassium feldspar   37 ____

38. If bedrock at a collisional plate boundary contains andalusite crystals, these crystals are changed into sillimanite and/or kyanite as temperature and pressure conditions increase. What is this process called?
(1) weathering (2) solidification (3) metamorphism (4) cementation 38 ____

Base your answers to questions 39 through 41 on the accompanying diagram, which has lettered arrows showing the motions of Earth and the Moon.

Key	
Arrow	**Motion**
A	Earth's rotation on its axis
B	Earth's revolution around the Sun
C	The Moon's rotation on its axis
D	The Moon's revolution around Earth

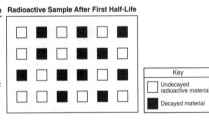

(Not drawn to scale)

39. These lettered arrows represent motions that are
(1) noncyclic and unpredictable
(2) noncyclic and predictable
(3) cyclic and unpredictable
(4) cyclic and predictable

39 _____

40. Which two motions are completed in about the same amount of time?
(1) A and B   (2) B and C   (3) C and D   (4) A and D   40 _____

41. Which lettered arrow represents the motion that causes the Moon to show phases when viewed from Earth?
(1) A   (2) B   (3) C   (4) D   41 _____

Base your answers to questions 42 and 43 on the accompanying diagram, which represents a model of a radio-active sample with a half-life of 5000 years. The white boxes represent undecayed radioactive material and the shaded boxes represent the decayed material after the first half-life.

Radioactive Sample After First Half-Life

Key	
☐	Undecayed radioactive material
■	Decayed material

42. How many *more* boxes should be shaded to represent the additional decayed material formed during the second half-life?
(1) 12   (2) 6   (3) 3   (4) 0   42 _____

43. Which radioactive isotope has a half-life closest in duration to this radioactive sample?
(1) carbon-14   (2) potassium-40   (3) uranium-238   (4) rubidium-87   43 _____

44. The photograph shows a sign near the Esopus Creek in Kingston, New York. The main purpose of the word "watershed" on this sign is to communicate that the Esopus Creek   (1) is a tributary of the Hudson River
(2) is a flood hazard where it flows into the Hudson River
(3) forms a delta in the Hudson River
(4) contains ancient fish fossils

44 _____

Base your answers to questions 45 and 46 on the accompanying diagrams. Diagrams *A*, *B*, and *C* represent three different river valleys.

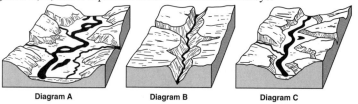

Diagram A   Diagram B   Diagram C

45. Which bar graph best represents the relative gradients of the main rivers shown in diagrams *A*, *B*, and *C*?

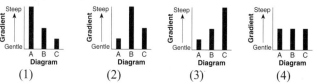

(1)   (2)   (3)   (4)   45 ____

46. Most sediments found on the floodplain shown in diagram *A* are likely to be
(1) angular and weathered from underlying bedrock
(2) angular and weathered from bedrock upstream
(3) rounded and weathered from underlying bedrock
(4) rounded and weathered from bedrock upstream   46 ____

Base your answers to questions 47 through 49 on the graph, which shows the amount of insolation during one year at four different latitudes on Earth's surface.

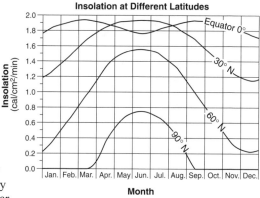

Insolation at Different Latitudes

47. This graph shows that insolation varies with
(1) latitude and time of day
(2) latitude and time of year
(3) longitude and time of day
(4) longitude and time of year   47 ____

48. Why is less insolation received at the equator in June than in March or September?
(1) The daylight period is longest at the equator in June.
(2) Winds blow insolation away from the equator in June.
(3) The Sun's vertical rays are north of the equator in June.
(4) Thick clouds block the Sun's vertical rays at the equator in June.   48 ____

49. Why is insolation 0 cal/cm²/min from October through February at 90° N?
(1) Intense cold prevents insolation from being absorbed during that time.
(2) Snowfields reflect sunlight during that time.
(3) The Sun is continually below the horizon during that time.
(4) Dust in the atmosphere blocks sunlight during that time.   49 ____

50. The diagram below shows tubes *A* and *B* partly filled with equal volumes of round plastic beads of uniform size. The beads in tube *A* are smaller than the beads in tube *B*. Water was placed in tube *A* until the pore spaces were filled. The drain valve was then opened, and the amount of time for the water to drain from the tube was recorded. The amount of water that remained around the beads was then calculated and recorded. Data table 1 shows the measurements recorded using tube *A*.

Data Table 1: Tube A	
water required to fill pore spaces	124 mL
time required for draining	2.1 sec
water that remained around the beads after draining	36 mL

If the same procedure was followed with tube *B*, which data table shows the measurements most likely recorded?

Data Table 2: Tube B	
water required to fill pore spaces	124 mL
time required for draining	1.4 sec
water that remained around the beads after draining	26 mL

(1)

Data Table 2: Tube B	
water required to fill pore spaces	124 mL
time required for draining	3.2 sec
water that remained around the beads after draining	36 mL

(3)

Data Table 2: Tube B	
water required to fill pore spaces	168 mL
time required for draining	3.2 sec
water that remained around the beads after draining	46 mL

(2)

Data Table 2: Tube B	
water required to fill pore spaces	168 mL
time required for draining	1.4 sec
water that remained around the beads after draining	36 mL

(4)

50 ____

Part B–2
Answer all questions in this part.

Directions (51–64): Record your answers in the spaces provided. Some questions may require the use of the *Earth Science Reference Tables.*

Base your answers to questions 51 and 52 on the diagrams below. The top diagram shows a depression and hill on a gently sloping area. The bottom diagram is a topographic map of the same area. Points *A*, *X*, and *Y* are locations on Earth's surface. A dashed line connects points *X* and *Y*. Elevation is indicated in feet.

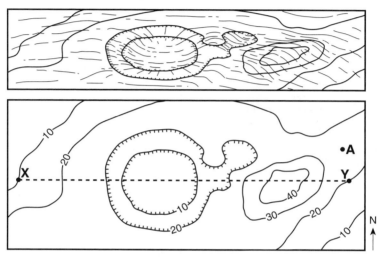

51. What is a possible elevation of point *A*? [1] _____ ft

52. On the grid below, construct a topographic profile along line *XY*, by plotting a point for the elevation of *each* contour line that crosses line *XY*. Points *X* and *Y* have already been plotted on the grid. Connect the points with a smooth, curved line to complete the profile. [2]

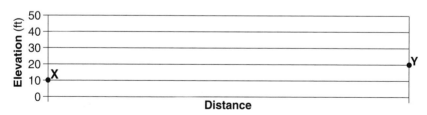

Base your answers to questions 53 through 55 on the flowchart below and on your knowledge of Earth science. The flowchart shows the formation of some igneous rocks. The circled letters *A*, *B*, *C*, and *D* indicate parts of the flowchart that have not been labeled.

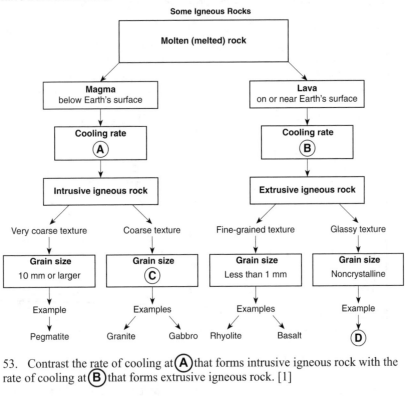

53. Contrast the rate of cooling at (A) that forms intrusive igneous rock with the rate of cooling at (B) that forms extrusive igneous rock. [1]

_____

_____

_____

_____

54. Give the numerical grain-size range that should be placed in the flowchart at (C).Units must be included in your answer. [1]

_____   to   _____

55. State *one* igneous rock that could be placed in the flowchart at (D). [1]

_____

Base your answers to questions 56 through 60 on the two diagrams below. Diagram I shows the orbits of the four inner planets. Black dots in diagram I show the positions in the orbits where each planet is closest to the Sun. Diagram II shows the orbits of the six planets that are farthest from the Sun. The distance scale in diagram II is different than the distance scale in diagram I.

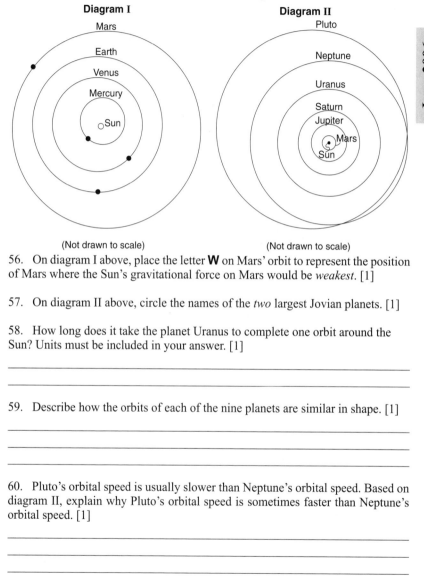

(Not drawn to scale)   (Not drawn to scale)

56. On diagram I above, place the letter **W** on Mars' orbit to represent the position of Mars where the Sun's gravitational force on Mars would be *weakest*. [1]

57. On diagram II above, circle the names of the *two* largest Jovian planets. [1]

58. How long does it take the planet Uranus to complete one orbit around the Sun? Units must be included in your answer. [1]

_____

_____

59. Describe how the orbits of each of the nine planets are similar in shape. [1]

_____

_____

_____

60. Pluto's orbital speed is usually slower than Neptune's orbital speed. Based on diagram II, explain why Pluto's orbital speed is sometimes faster than Neptune's orbital speed. [1]

_____

_____

_____

Base your answers to questions 61 through 64 on the map below, which shows the different lobes (sections) of the Laurentide Ice Sheet, the last continental ice sheet that covered most of New York State. The arrows show the direction that the ice lobes flowed. The terminal moraine shows the maximum advance of this ice sheet.

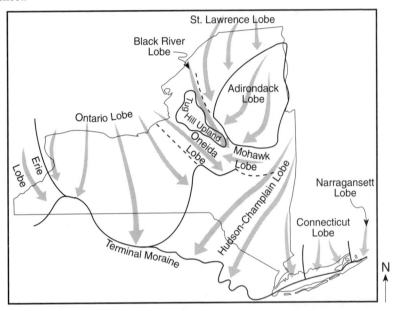

61. During which geologic epoch did the Laurentide Ice Sheet advance over New York State? [1] _____Epoch

62. Describe the arrangement of rock material in the sediments that were directly deposited by the glacier. [1]

_____

_____

_____

63. According to the map, toward which compass direction did the ice lobe flow over the Catskills? [1] _____

64. What evidence might be found on surface bedrock of the Catskills that would indicate the direction of ice flow in this region? [1]

_____

_____

_____

Part C
Answer all questions in this part.

Directions (65–83): **Record your answers in the spaces provided. Some questions may require the use of the *Earth Science Reference Tables*.**

Base your answers to questions 65 and 66 on the accompanying diagram, which represents water molecules attached to salt and dust particles within a cloud in the atmosphere.

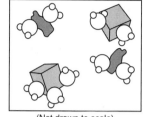

(Not drawn to scale)

65. Explain why salt and dust particles are important in cloud formation. [1]

_____

_____

_____

66. State *one* natural process that causes large amounts of dust to enter Earth's atmosphere. [1]

_____

Base your answers to questions 67 through 69 on the cross section and bar graph below. The cross section shows a portion of Earth's crust along the western coast of the United States. The points show different locations on Earth's surface. The arrows show the prevailing wind direction. The bar below each point shows the yearly precipitation at that location.

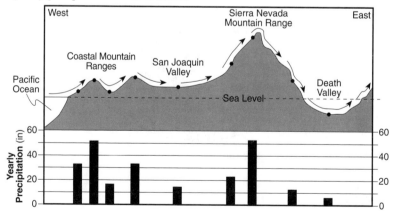

67. Explain why the valleys have *lower* amounts of precipitation than points on the western slopes of the mountain ranges. [1]

_____

_____

_____

_____

68. What is the yearly precipitation total for the four points located in the Coastal Mountain Ranges? [1] _____in

69. State *one* reason why colder temperatures would be recorded at the top of the Sierra Nevada Mountain Range than at the top of the Coastal Mountain Ranges. [1]

_____

_____

_____

Base your answers to questions 70 and 71 on the diagram below, which shows the latitude-longitude grid on a model of Earth. Point Y is a location on Earth's surface.

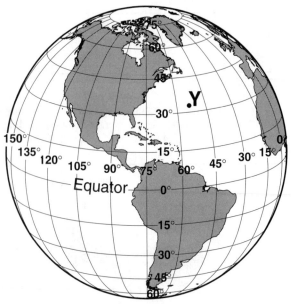

70. On the diagram above, place an **X** at 15° S 30° W. [1]

71. What is Earth's rate of rotation at point Y, in degrees per hour? [1] _____ °/hr

Base your answers to questions 72 through 76 on the two maps below and on your knowledge of Earth science. Both maps show data from a December snowstorm. Map 1 shows the snowfall, measured in inches, at various locations in New York State, Pennsylvania, and New Jersey. Map 2 shows weather conditions in New York State and the surrounding region during the storm. Letter **L** represents the center of the low-pressure system that produced the snowstorm. Isobars show air pressure, in millibars.

Map 1
December Snowfall Amounts (inches)

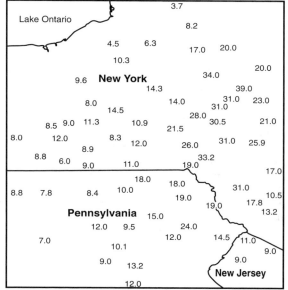

Map 2

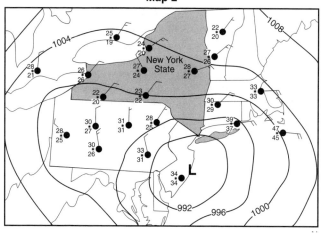

72. On map 1 on the previous page, draw the 30.0-inch snowfall insoline. Assume that the decimal point for each snowfall depth marks the exact location where the snowfall was measured. [1]

73. Most residents knew this storm was coming. State *one* action a New York State resident should have taken to prepare for a snow emergency. [1]

_____

_____

_____

_____

74. Using map 2 on the previous page, complete the table below by describing the weather conditions at Buffalo, New York. [2]

Weather Conditions	Description
present weather	
wind direction from	
wind speed (knots)	
relative humidity (%)	

75. Describe the general surface wind pattern around the low-pressure center shown on map 2. [1]

_____

_____

_____

_____

76. Toward which compass direction would this low-pressure center most likely have moved if this system followed a normal storm track? [1]

_____

Base your answers to questions 77 through 80 on the geologic cross section. The rock layers have not been overturned. Point *A* is located in the zone of contact metamorphism.

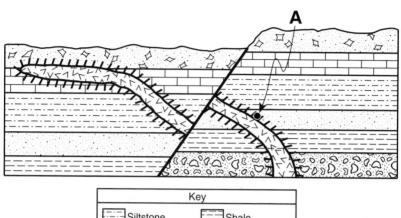

Key

Siltstone		Shale	
Limestone		Basalt intrusion	
Sandstone		Breccia	
Conglomerate		Contact metamorphism	

77. Which metamorphic rock most likely formed at point *A*? [1]

_____

78. State the evidence shown by the cross section that supports the inference that the fault is younger than the basalt intrusion. [1]

_____

_____

_____

79. List basalt, limestone, and breccia in the order in which they were formed. [1]

Formed first: _____

Formed second: _____

Formed last: _____

80. What is the largest silt particle that could be found in the siltstone layer? [1] _____ **cm**

Base your answers to questions 81 through 83 on the passage below and on your knowledge of Earth science.

A New Oregon Volcano?

The Three Sisters are 10,000-foot volcanic mountain peaks in Oregon. Volcanic eruptions began building the Three Sisters from andesitic lava and cinders 700,000 years ago. The last major eruption occurred 2000 years ago.

West of the Three Sisters peaks, geologists have recently discovered that Earth's surface is bulging upward in a bull's-eye pattern 10 miles wide. There is a 4-inch rise at its center, which geologists believe could be the beginning of another volcano. The uplift was found by comparing satellite images. This uplift in Oregon may allow the tracking of a volcanic eruption from its beginning, long before the smoke and explosions begin.

This uplift is most likely caused by an upflow of molten rock from more than four miles below the surface. Rock melts within Earth's interior and then moves upward in cracks in Earth's crust, where it forms large underground pools called magma chambers. Magma upwelling often produces signs that help scientists predict eruptions and protect humans. When the pressure of rising magma becomes forceful enough to crack bedrock, swarms of small earthquakes occur. Rising magma releases carbon dioxide and other gases that can be detected at the surface.

81. Identify *one* of the minerals found in the andesite rock of the Three Sisters volcanoes. [1] _____

82. The cross section below represents Earth's interior beneath the Three Sisters. Place a triangle, ▲, on the cross section to indicate the location where the new volcano will most likely form. [1]

83. On the same cross section below, place arrows through each point, *X*, *Y*, and *Z*, to indicate the relative motion of *each* of these sections of the lithosphere. [1]

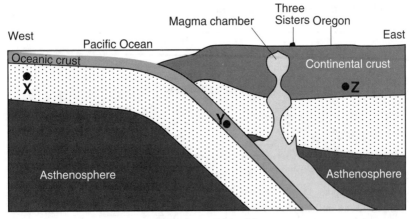

(Not drawn to scale)

Directions (1–35): For *each* statement or question, write in the space provided the *number* of the word or expression that, of those given, best completes the statement or answers the question. Some questions may require the use of the *Earth Science Reference Tables.*

1. The best evidence that Earth spins on its axis is the motion of
(1) tectonic plates   (3) a wind vane
(2) *Polaris*   (4) a Foucault pendulum   1 ____

2. When viewed from Earth, the light from very distant galaxies shows a red shift. This is evidence that these distant galaxies are
(1) revolving around the Sun
(2) revolving around the Milky Way
(3) moving away from Earth
(4) moving toward Earth   2 ____

3. The arrows on the accompanying cross section show the prevailing wind that flows over a mountain. Points *A* and *B* represent locations on opposite sides of the mountain. Which statement correctly describes the differences in the climates of locations *A* and *B*?

(1) Location *A* is warmer and drier than location *B*.
(2) Location *A* is cooler and wetter than location *B*.
(3) Location *B* is warmer and wetter than location *A*.
(4) Location *B* is cooler and drier than location *A*.   3 ____

4. The average temperature at Earth's equator is higher than the average temperature at Earth's South Pole because the South Pole
(1) receives less intense insolation
(2) receives more infrared radiation
(3) has less land area
(4) has more cloud cover   4 ____

5. Which statement best summarizes the general effects of ocean currents at 20° S latitude on coastal regions of South America?
(1) The east coast and west coast are both warmed.
(2) The east coast and west coast are both cooled.
(3) The east coast is warmed and the west coast is cooled.
(4) The east coast is cooled and the west coast is warmed.   5 _____

6. Which type of electromagnetic energy has the longest wavelength?
(1) infrared radiation   (3) ultraviolet radiation
(2) radio wave radiation   (4) x-ray radiation   6 _____

7. Under which atmospheric conditions will water most likely evaporate at the fastest rate?
(1) hot, humid, and calm   (3) cold, humid, and windy
(2) hot, dry, and windy   (4) cold, dry, and calm   7 _____

8. Which temperature zone of Earth's atmosphere contains the most water vapor?
(1) mesosphere   (3) thermosphere
(2) stratosphere   (4) troposphere   8 _____

9. The accompanying diagram shows the result of leaving an empty, dry clay flowerpot in a full container of water for a period of time. The water level in the container dropped to level A. The top of the wet area moved to level B. Level B is higher than level A because water

B (wet to this height)
A (water level)
Container

(1) is less dense than the clay pot
(2) is more dense than the clay pot
(3) traveled upward in the clay pot by capillary action
(4) traveled downward in the clay pot by capillary action   9 _____

10. Which weather condition most directly determines wind speeds at Earth's surface?
(1) visibility changes   (3) air-pressure gradient
(2) amount of cloud cover   (4) dewpoint differences   10 _____

11. Which statement best explains why an increase in the relative humidity of a parcel of air generally increases the chance of precipitation?
(1) The dewpoint is farther from the condensation point, causing rain.
(2) The air temperature is closer to the dewpoint, making cloud formation more likely.
(3) The amount of moisture in the air is greater, making the air heavier.
(4) The specific heat of the moist air is greater than the drier air, releasing energy.   11 _____

12. The two elements that make up the largest percentage by mass of Earth's crust are oxygen and
(1) silicon   (2) potassium   (3) hydrogen   (4) nitrogen   12 _____

13. The accompanying weather instrument can be used to determine relative humidity. Based on the temperatures shown, the relative humidity is

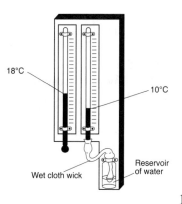

18°C

10°C

Wet cloth wick

Reservoir of water

(1) 19%
(2) 2%
(3) 33%
(4) 40%   13 _____

14. Which two minerals have cleavage planes at right angles?
(1) biotite mica and muscovite mica
(2) sulfur and amphibole
(3) quartz and calcite
(4) halite and pyroxene   14 _____

15. Which property would best distinguish sediment deposited by a river from sediment deposited by a glacier?
(1) mineral composition of the sediment
(2) amount of sediment sorting
(3) thickness of sediment layers
(4) age of fossils found in the sediment   15 _____

16. Salt deposits are found in the surface bedrock near which
New York State location?
(1) Oswego   (2) Syracuse   (3) Old Forge   (4) Albany   16 _____

17. The photograph below shows a sand dune that formed in a coastal area.

This sand dune was most likely formed by
(1) water flowing from the left   (3) wind blowing from the left
(2) water flowing from the right (4) wind blowing from the right 17 _____

18. What is the origin of fine-grained igneous rock?
(1) lava that cooled slowly on Earth's surface
(2) lava that cooled quickly on Earth's surface
(3) silt that settled slowly in ocean water
(4) silt that settled quickly in ocean water   18 _____

19. Why does the oceanic crust sink beneath the continental crust
at a subduction boundary?
(1) The oceanic crust has a greater density.
(2) The oceanic crust is pulled downward by Earth's magnetic field.
(3) The continental crust has a more mafic composition.
(4) The continental crust is pulled upward by the Moon's gravity. 19 _____

20. Based on fossil evidence, most scientists infer that
(1) life has not changed significantly throughout Earth's history
(2) life has evolved from complex to simple forms
(3) many organisms that lived on Earth have become extinct
(4) mammals developed early in the Precambrian time period   20 _____

21. The presence of which index fossil in the surface bedrock
most likely indicates that a forest environment once existed
in the region?
(1) *Aneurophyton*   (3) *Centroceras*
(2) *Cystiphyllum*   (4) *Bothriolepis*   21 _____

22. The accompanying block diagram shows a displacement of rock layers. Which process describes the downward sliding of the rock material?
(1) tidal changes
(2) glacial erosion
(3) mass movement
(4) lava flow

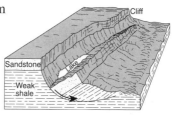

22____

23. Which two types of rock are most commonly found as outcrops in New York State's Newark Lowlands landscape region?
(1) rock salt and gypsum
(2) limestone and granite
(3) gneiss and quartzite
(4) conglomerate and sandstone   23____

24. Which processes most likely formed the shale bedrock found near Ithaca, New York?
(1) uplift and solidification
(2) burial and compaction
(3) heat and pressure
(4) melting and recrystallization   24____

25. The symbols below are used to represent different regions of space.

Universe = □   Earth = ○   Galaxy = ▱   Solar system = ○

Which diagram shows the correct relationship between these four regions? [If one symbol is within another symbol, that means it is part of, or included in, that symbol.]

(1)   (2)   (3)   (4)   25____

26. The diagram below shows four surfaces of equal area that absorb insolation.

A   B   C   D

Which letter represents the surface that most likely absorbs the greatest amount of insolation?
(1) A   (2) B   (3) C   (4) D   26____

27. A student in New York State looked toward the eastern horizon to observe sunrise at three different times during the year. The student drew the following diagram that shows the positions of sunrise, A, B, and C, during this one-year period.

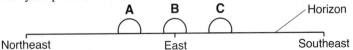

Which list correctly pairs the location of sunrise to the time of the year?

(1) A—June 21
    B—March 21
    C—December 21

(3) A—March 21
    B—June 21
    C—December 21

(2) A—December 21
    B—March 21
    C—June 21

(4) A—June 21
    B—December 21
    C—March 21

27 _____

28. Which graph best represents the average monthly temperatures for one year at a location in the Southern Hemisphere?

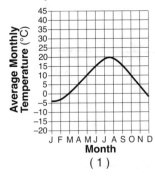

(1)

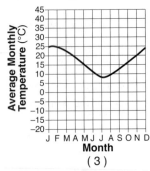

(3)

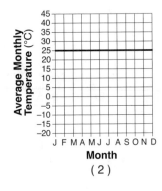

(2)

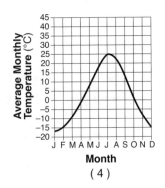

(4)

28 _____

29. Which station model shows the correct form for indicating a northwest wind at 25 knots and an air pressure of 1023.7 mb?

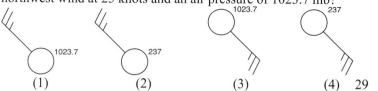

(1) (2) (3) (4) 29 ____

30. The data table below shows the mass and volume of three samples of the same mineral. [The density column is provided for student use.]

Data Table

Sample	Mass (g)	Volume (cm³)	Density (g/cm³)
A	50	25	
B	100	50	
C	150	75	

Which graph best represents the relationship between the density and the volume of these mineral samples?

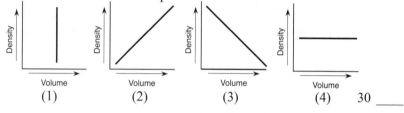

(1) (2) (3) (4) 30 ____

Base your answers to questions 31 and 32 on the accompanying geologic cross section. Location *A* is within the metamorphic rock.

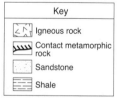

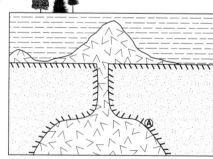

Key

⊿△	Igneous rock
〰〰	Contact metamorphic rock
· ·	Sandstone
≡≡	Shale

31. The metamorphic rock at location *A* is most likely
(1) marble (2) quartzite (3) phyllite (4) slate 31 ____

32. Which rock is the youngest?
(1) shale (2) sandstone (3) igneous rock (4) rock at location *A* 32 ____

Base your answers to questions 33 and 34 on the map below, which shows the risk of damage from seismic activity in the United States.

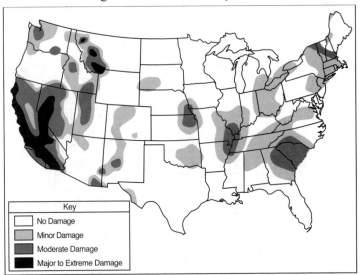

Key
No Damage
Minor Damage
Moderate Damage
Major to Extreme Damage

33. In the United States, most of the major damage expected from a future earthquake is predicted to occur near a
(1) divergent plate boundary, only
(2) convergent plate boundary, only
(3) mid-ocean ridge and a divergent plate boundary
(4) transform plate boundary and a hot spot   33 _____

34. Which New York State location has the greatest risk of earthquake damage?
(1) Binghamton (2) Buffalo   (3) Plattsburgh (4) Elmira   34 _____

35. Which pie graph best represents the percentage of total time for the four major divisions of geologic time?

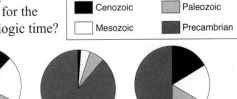

Key
Cenozoic   Paleozoic
Mesozoic   Precambrian

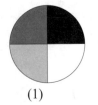   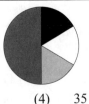

(1)   (2)   (3)   (4)   35 _____

Part B–1

Answer all questions in this part.

Directions (36–50): For *each* statement or question, write in the space provided the *number* of the word or expression that, of those given, best completes the statement or answers the question. Some questions may require the use of the *Earth Science Reference Tables.*

Base your answers to questions 36 through 38 on the world map below. Letters *A* through *D* represent locations on Earth's surface.

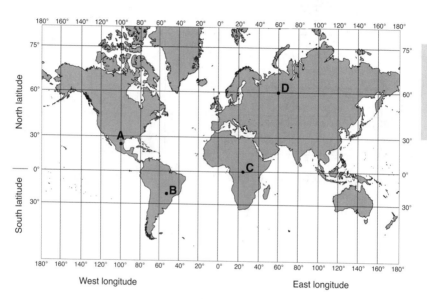

36. At which location could an observer *not* see *Polaris* in the night sky at any time during the year?

(1) *A*   (2) *B*   (3) *C*   (4) *D*   36 ____

37. Which location receives 12 hours of daylight and 12 hours of darkness on June 21?

(1) *A*   (2) *B*   (3) *C*   (4) *D*   37 ____

38. At which location on December 21 is the Sun directly overhead at solar noon?

(1) *A*   (2) *B*   (3) *C*   (4) *D*   38 ____

Base your answers to questions 39 through 42 on the table below, which shows eight inferred stages describing the formation of the universe from its beginning to the present time.

Data Table

Stage	Description of the Universe	Average Temperature of the Universe (°C)	Time From the Beginning of Universe
1	the size of an atom	?	0 second
2	the size of a grapefruit	?	10^{-43} second
3	"hot soup" of electrons	10^{27}	10^{-32} second
4	Cooling allows protons and neutrons to form.	10^{13}	10^{-6} second
5	still too hot to allow the forming of atoms	10^{8}	3 minutes
6	Electrons combine with protons and neutrons, forming hydrogen and helium atoms. Light emission begins.	10,000	300,000 years
7	Hydrogen and helium form giant clouds (nebulae) that will become galaxies. First stars form.	−200	1 billion years
8	Galaxy clusters form and first stars die. Heavy elements are thrown into space, forming new stars and planets.	−270	13.7 billion years

39. How soon did protons and neutrons form after the beginning of the universe?
(1) 10^{-43} second   (3) 10^{-6} second
(2) 10^{-32} second   (4) 13.7 billion years   39 _____

40. What is the most appropriate title for this table?
(1) The Big Bang Theory
(2) The Theory of Plate Tectonics
(3) The Law of Superposition
(4) The Laws of Planetary Motion   40 _____

41. According to this table, the average temperature of the universe since stage 3 has
(1) decreased, only   (3) remained the same
(2) increased, only   (4) increased, then decreased   41 _____

42. Between which two stages did our solar system form?
(1) 1 and 3   (2) 3 and 5   (3) 6 and 7   (4) 7 and 8   42 _____

Base your answers to questions 43 and 44 on the map below, which shows Earth's Southern Hemisphere and the inferred tectonic movement of the continent of Australia over geologic time. The arrows between the dots show the relative movement of the center of the continent of Australia. The parallels of latitude from 0° to 90° south are labeled.

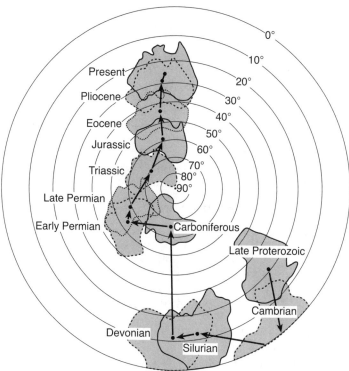

43. The geographic position of Australia on Earth's surface has been changing mainly because
(1) the gravitational force of the Moon has been pulling on Earth's landmasses
(2) heat energy has been creating convection currents in Earth's interior
(3) Earth's rotation has spun Australia into different locations
(4) the tilt of Earth's axis has changed several times   43 ___

44. During which geologic time interval did Australia most likely have a warm, tropical climate because of its location?
(1) Cambrian   (3) Late Permian
(2) Carboniferous   (4) Eocene   44 ___

Base your answers to questions 45 through 47 on the accompanying map, which shows Earth's planetary wind belts.

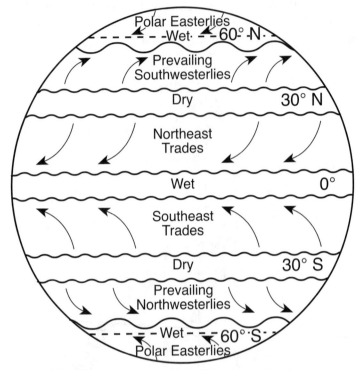

45. The curving of these planetary winds is the result of
(1) Earth's rotation on its axis
(2) the unequal heating of Earth's atmosphere
(3) the unequal heating of Earth's surface
(4) Earth's gravitational pull on the Moon   45 ____

46. Which wind belt has the greatest effect on the climate of New York State?
(1) prevailing northwesterlies   (3) northeast trades
(2) prevailing southwesterlies   (4) southeast trades   46 ____

47. Which climatic conditions exist where the trade winds converge?
(1) cool and wet   (3) warm and wet
(2) cool and dry   (4) warm and dry   47 ____

Base your answers to questions 48 through 50 on the diagram below, which shows a model used to investigate the erosional-depositional system of a stream. The model was tilted to create a gentle slope, and a hose supplied water to form the meandering stream shown.

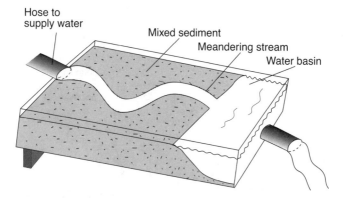

48. Which diagram best represents where erosion, E, and deposition, D, are most likely occurring along the curves of the meandering stream?

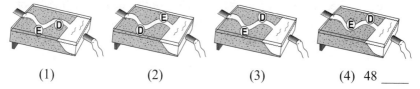

(1)   (2)   (3)   (4) 48 _____

49. Which diagram best represents the arrangement of large, L, and small, S, sediment deposited as the stream enters the water basin?

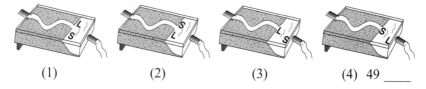

(1)   (2)   (3)   (4) 49 _____

50. How can the model be changed to increase the amount of sediment transported by the stream?
(1) decrease the temperature of the sediment
(2) decrease the slope
(3) increase the size of the sediment
(4) increase the rate of the water flow   50 _____

Part B–2

Answer all questions in this part.

Directions (51–64): Record your answers in the spaces provided. Some questions may require the use of the *Earth Science Reference Tables.*

Base your answers to questions 51 through 53 on the accompanying weather map. The weather map shows a low-pressure system in New York State during July. The **L** represents the center of the low-pressure system. Two fronts extend from the center of the low. Line *XY* on the map is a reference line.

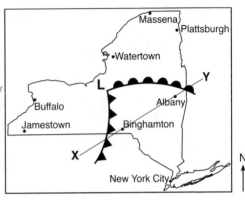

51. The cross section below shows a side view of the area along line *XY* on the map. On lines 1 and 2 in the cross section, place the appropriate two-letter air-mass symbols to identify the most likely type of air mass at *each* of these locations. [1]

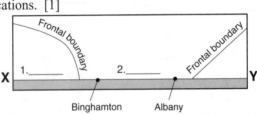

52. The forecast for one city located on the map is given below:

"In the next hour, skies will become cloud covered. Heavy rains are expected with possible lightning and thunder. Temperatures will become much cooler."

State the name of the city for which this forecast was given. [1]

_____

53. Identify *one* action that people should take to protect themselves from lightning. [1]

_____

Base your answers to questions 54 through 57 on the diagrams below, which represent two bedrock outcrops, I and II, found several kilometers apart in New York State. Rock layers are lettered *A* through *F*. Drawings represent specific index fossils.

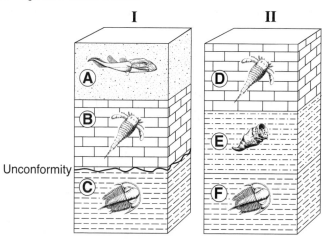

54. During which geologic time period was rock layer *C* deposited? [1] _____ Period

55. Identify *two* processes that produced the unconformity in outcrop I. [1]

(1) _____   (2) _____

56. Describe *one* characteristic a fossil must have in order to be considered a good index fossil. [1]

_____

57. Explain why carbon-14 can *not* be used to find the geologic age of these index fossils. [1]

_____

_____

_____

Base your answers to questions 58 through 61 on the cross section and block diagram below. The cross section shows an enlarged view of the stream shown in the block diagram. The sediments in the cross section are drawn to actual size. Arrows show the movement of particles in the stream. The block diagram represents a region of Earth's surface and the bedrock beneath the region.

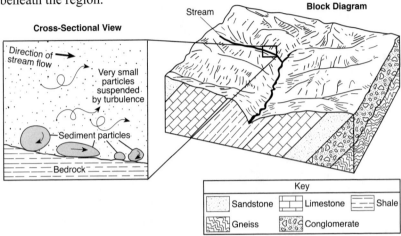

Key

Sandstone Limestone Shale
Gneiss Conglomerate

58. After measuring the actual size, identify the name of the largest particle shown on the stream bottom in the cross section. [1]

_____

59. What process is responsible for producing the rounded shape of the particles shown on the stream bottom in the cross section? [1]

_____

60. Identify the type of rock shown in the block diagram that appears to be the most easily eroded. [1]

_____

61. How does the shape of a valley eroded by a glacier differ from the shape of the valley shown in the block diagram? [1]

_____

_____

Base your answers to questions 62 through 64 on the accompanying photograph of a sample of gneiss.

62. What observable characteristic could be used to identify this rock sample as gneiss? [1]

_____

63. Identify *two* minerals found in gneiss that contain iron and magnesium. [1]

_____ and _____

64. A dark-red mineral with a glassy luster was also observed in this gneiss sample. Identify the mineral and state *one* possible use for this mineral. [1]

Mineral:_____

Use:_____

Part C
Answer all questions in this part.

Directions (65–82): **Record your answers in the spaces provided. Some questions may require the use of the *Earth Science Reference Tables*.**

Base your answers to questions 65 through 69 on the accompanying data table, which shows the percentage of the lighted side of the Moon visible from Earth for the first fourteen days of July 2003.

65. On what July date listed in the table did the Moon appear as shown below? [1]

July _____

Date	Percentage of Lighted Side of the Moon Visible From Earth (%)
July 1	1
July 2	5
July 3	10
July 4	17
July 5	26
July 6	37
July 7	48
July 8	59
July 9	70
July 10	80
July 11	89
July 12	95
July 13	98
July 14	100

66. What motion of the Moon causes the percentage of the lighted side of the Moon visible from Earth to change from July 1 to July 14? [1]

_____

67. A full Moon phase was observed on July 14. On what day in August was the next full Moon phase observed? [1]

August _____

68. The accompanying diagram shows the orbit of the Moon around Earth. Place an **X** on the orbit to show where the Moon was in its orbit on July 14, 2003. [1]

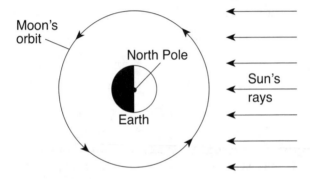

69. Why are the phases of the Moon considered to be cyclic? [1]

_____

_____

_____

_____

_____

Base your answers to questions 70 through 73 on the topographic map shown below. Letters *A*, *B*, *C*, *D*, and *E* represent locations on Earth's surface. Letters *K*, *L*, *M*, and *N* are locations along Copper Creek. Elevations are measured in meters.

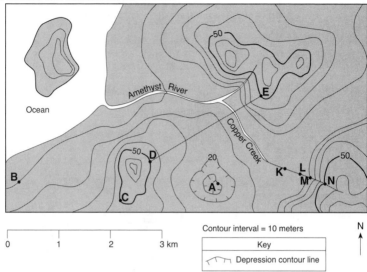

Contour interval = 10 meters

Key

⌢⌢⌢ Depression contour line

0   1   2   3 km

N ↑

70. What is the elevation of location *A*? [1] _____ m

71. Calculate the gradient between points *B* and *C* and label your answer with the correct units. [2] Gradient = _____

72. On the accompanying grid, construct a topographic profile along line *DE* by plotting an **X** for the elevation of each contour line that crosses line *DE*. Connect the **X**s with a smooth, curved line to complete the profile. [2]

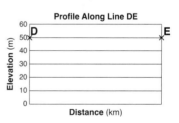

Profile Along Line DE

73. Explain how the map indicates that Copper Creek flows faster between points *N* and *M* than between points *L* and *K*. [1]

_____

_____

Base your answers to questions 74 through 76 on the example of a seismogram and set of instructions for determining the Richter magnitude of an earthquake below. The example shows the Richter magnitude of an earthquake 210 kilometers from a seismic station.

Example of a Seismogram of an Earthquake

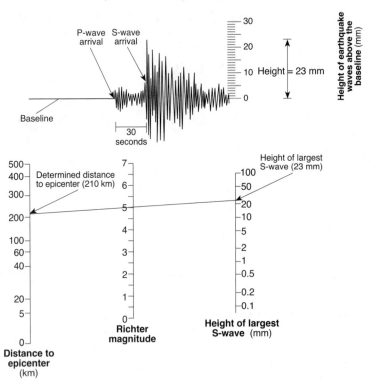

Instructions for determining Richter magnitude:
- Determine the distance to the epicenter of the earthquake. (The distance in the example is 210 kilometers.)
- Measure the maximum wave height of the S-wave recorded on the seismogram. (The height in the example is 23 millimeters.)
- Place a straightedge between the distance to the epicenter (210 kilometers) and the height of the largest S-wave (23 millimeters) on the appropriate scales. Draw a line connecting these two points. The magnitude of the earthquake is determined by where the line intersects the Richter magnitude scale. (The magnitude of this example is 5.0.)

74. Using the set of instructions the previous page and the seismogram and scales below, determine the Richter magnitude of an earthquake that was located 500 kilometers from this seismic station. [1]

Seismogram of an Earthquake

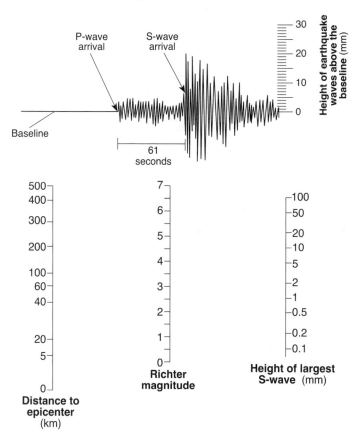

74 _____

75. Identify the information shown on the seismogram that was used to determine that the distance to the epicenter was 500 kilometers. [1]

_____

_____

76. How long did it take the first S-wave to travel 500 kilometers to reach this seismic station? [1]

_____minutes _____ seconds

Base your answers to questions 77 through 81 on the passage below and on the map on the next page. The passage describes the Gakkel Ridge found at the bottom of the Arctic Ocean. The map shows the location of he Gakkel Ridge.

The Gakkel Ridge

In the summer of 2001, scientists aboard the U.S. Coast Guard icebreaker *Healy* visited one of the least explored places on Earth. The scientists studied the 1800-kilometer-long Gakkel Ridge at the bottom of the Arctic Ocean near the North Pole. The Gakkel Ridge is a section of the Arctic Mid-Ocean Ridge and extends from the northern end of Greenland across the Arctic Ocean floor toward Russia. At a depth of about 5 kilometers below the ocean surface, the Gakkel Ridge is one of the deepest mid-ocean ridges in the world. The ridge is believed to extend down to Earth's mantle, and the new seafloor being formed at the ridge is most likely composed of huge slabs of mantle rock. Bed-rock samples taken from the seafloor at the ridge were determined to be the igneous rock peridotite.

The Gakkel Ridge is also the slowest moving mid-ocean ridge. Some ridge systems, like the East Pacific Ridge, are rifting at a rate of about 20 centimeters per year. The Gakkel Ridge is rifting at an average rate of less than 1 centimeter per year. This slow rate of move-ment means that there is less volcanic activity along the Gakkel Ridge than along other ridge systems. However, heat from the underground magma slowly seeps up through cracks in the rocks of the ridge at structures scientists call hydrothermal (hot water) vents. During the 2001 cruise, a major hydrothermal vent was discovered at 87° N latitude 45° E longitude.

77. On the map on the next page, place an **X** on the location of the major hydrothermal vent described in the passage. [1]

78. Describe the relative motion of the two tectonic plates on either side of the Gakkel Ridge. [1]

_____

_____

79. The Gakkel Ridge is a boundary between which two tectonic plates? [1]
_____ Plate and _____ Plate

Diagram for question 77.

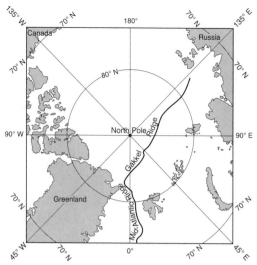

80. Identify *one* feature, other than hydrothermal vents, often found at mid-ocean ridges like the Gakkel Ridge that indicates heat from Earth's interior is escaping. [1]

_____

_____

81. State the *two* minerals that were most likely found in the igneous bedrock samples collected at the Gakkel Ridge. [1]

_____ and _____

82. The diagram below shows a view of the ground from directly above a flagpole in New York State at solar noon on a particular day of the year. The flagpole's shadow at solar noon is shown. Draw the position and relative length of the shadow that would be cast by this flagpole three hours later. [2]

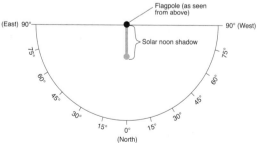

Part A

Answer all questions in this part.

Directions (1-35): For *each* statement or question, write in the space provided the number of the word or expression that, of those given, best completes the statement or answers the question. Some questions may require the use of the *Earth Science Reference Tables.*

1. As viewed from Earth, most stars appear to move across the sky each night because
(1) Earth revolves around the Sun
(2) Earth rotates on its axis
(3) stars orbit around Earth
(4) stars revolve around the center of the galaxy   1 _____

2. The star *Algol* is estimated to have approximately the same luminosity as the star *Aldebaran* and approximately the same temperature as the star *Rigel*. *Algol* is best classified as a
(1) main sequence star   (3) white dwarf star
(2) red giant star   (4) red dwarf star   2 _____

3. The explosion associated with the Big Bang theory and the formation of the universe is inferred to have occurred how many billion years ago?
(1) less than 1   (2) 2.5   (3) 4.6   (4) over 10   3 _____

4. The accompanying diagram represents the constellation Lyra. Which statement best explains why Lyra is visible to an observer in New York State at midnight in July but *not* visible at midnight in December?
(1) Earth spins on its axis,   (3) Lyra spins on its axis.
(2) Earth orbits the Sun.   (4) Lyra orbits Earth.   4 _____

5. The Coriolis effect provides evidence that Earth
(1) rotates on its axis   (3) undergoes cyclic tidal changes
(2) revolves around the Sun   (4) has a slightly eccentric orbit   5 _____

6. The altitude of the ozone layer near the South Pole is 20 kilometers above sea level. Which temperature zone of the atmosphere contains this ozone layer?
(1) troposphere   (2) stratosphere (3) mesosphere (4) thermosphere   6 _____

7. Air masses are identified on the basis of temperature and
(1) type of precipitation   (3) moisture content
(2) wind velocity   (4) atmospheric transparency   7 _____

8. A low-pressure system in the Northern Hemisphere has a surface air-circulation pattern that is
(1) clockwise and away from the center
(2) clockwise and toward the center
(3) counterclockwise and away from the center
(4) counterclockwise and toward the center   8 _____

9. During some winters in the Finger Lakes region of New York State, the lake water remains unfrozen even though the land around the lakes is frozen and covered with snow. The primary cause of this difference is that water
(1) gains heat during evaporation   (3) has a higher specific heat
(2) is at a lower elevation   (4) reflects more radiation   9 _____

10. The reaction below represents an energy-producing process.

Hydrogen   +   Hydrogen   →   Helium   + Energy
(lighter element)   (lighter element)   (heavier element)

The reaction represents how energy is produced
(1) in the Sun by fusion
(2) when water condenses in Earth's atmosphere
(3) from the movement of crustal plates
(4) during nuclear decay   10 _____

11. Diagram 1 shows the Moon in its orbit at four positions labeled *A*, *B*, *C*, and *D*. Diagram 2 shows a phase of the Moon as viewed from New York State.

At which labeled Moon position would the phase of the Moon shown in diagram 2 be observed from New York State?
(1) *A*   (2) *B*   (3) *C*   (4) *D*   11 _____

12. The accompanying diagram shows an observer measuring the altitude of *Polaris*. What is the latitude of the observer?
(1) 20° N
(2) 20° S
(3) 70° N
(4) 70° S   12 _____

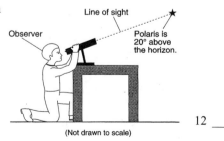

13. The accompanying diagram shows the spectral lines for an element.

 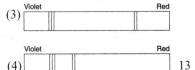

Which diagram best represents the spectral lines of this element when its light is observed coming from a star that is moving away from Earth?

(1) Violet ——————— Red

(3) Violet ——————— Red

(2) Violet ——————— Red

(4) Violet ——————— Red

13 _____

14. The accompanying weather instrument shown can be used to determine dewpoint. Based on the values shown, the dewpoint is
(1) –5°C
(2) 2°C
(3) 8°C
(4) 33°C

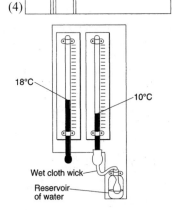

14 _____

15. Which arrangement of the Sun, the Moon, and Earth results in the highest high tides, and the lowest low tides on Earth? (Diagrams are not drawn to scale.)

(1)   (2)   (3)   (4)   15 _____

16. Which station model correctly represents the weather conditions in an area that is experiencing winds from the northeast at 25 knots and has had a steady drop in barometric pressure of 2.7 millibars during the last three hours?

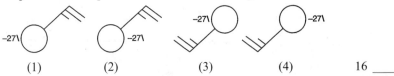

(1)   (2)   (3)   (4)   16 _____

17. Which ocean current carries cool water toward Earth's equator?
(1) Alaska Current   (3) Peru Current
(2) East Australia Current (4) North Atlantic Current   17 _____

18. The diagram shows weather instruments *A* and *B*. Which table correctly indicates the name of the weather instrument and the weather variable that it measures?

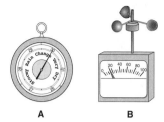

A   B

Instrument		Weather Variable Measured
Letter	Name	
A	thermometer	humidity
B	wind vane	wind direction

(1)

Instrument		Weather Variable Measured
Letter	Name	
A	barometer	wind speed
B	anemometer	air pressure

(3)

Instrument		Weather Variable Measured
Letter	Name	
A	thermometer	wind direction
B	wind vane	humidity

(2)

Instrument		Weather Variable Measured
Letter	Name	
A	barometer	air pressure
B	anemometer	wind speed

(4)   18 _____

19. Equal areas of which surface would most likely absorb the most insolation?
(1) smooth, white surface
(2) rough, white surface
(3) smooth, black surface
(4) rough, black surface   19 _____

20. On the map below, the darkened areas represent locations where living corals currently exist. The arrow points to a location where coral fossils have been found in Devonian-age bedrock in New York State.

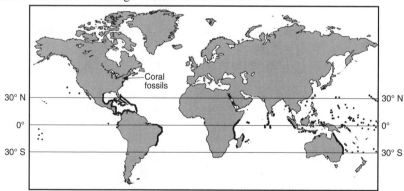

Devonian-age coral fossils found in some New York State bedrock are *not* located in the same general region that present-day corals are living because during the Devonian Period
(1) corals migrated to New York State
(2) corals lived everywhere on Earth
(3) New York State was closer to the equator
(4) New York State had a colder climate   20 _____

21. Which index fossil may be found in the surface bedrock near Ithaca, New York?

Elliptocephala (1) Coelophysis (2) Bothriolepis (3) Maclurites (4)

21 _____

22. The cross sections below represent three widely separated outcrops of exposed bedrock. Letters A, B, C, and D represent fossils found in the rock layers.

Which fossil appears to have the best characteristics of an index fossil?
(1) A   (2) B   (3) C   (4) D

22 _____

23. Active volcanoes are most abundant along the
(1) edges of tectonic plates   (3) 23.5° N and 23.5° S parallels of latitude
(2) eastern coastline of continents   (4) equatorial ocean floor

23 _____

24. Which part of Earth's interior is inferred to have convection currents that cause tectonic plates to move?
(1) rigid mantle   (2) asthenosphere (3) outer core (4) inner core

24 _____

25. Compared to the continental crust, the oceanic crust is
(1) less dense and less felsic   (3) more dense and more felsic
(2) less dense and less mafic   (4) more dense and more mafic

25 _____

26. The accompanying pie graph shows the elements comprising Earth's crust in percent by mass. Which element is represented by the letter **X**?
(1) silicon   (3) nitrogen
(2) lead   (4) hydrogen

26 _____

27. The block diagrams below show two landscape regions labeled *A* and *B*

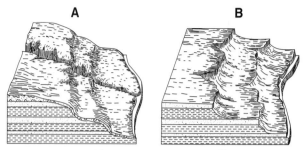

What is the most probable cause of the difference in surface features between *A* and *B*?

(1) *A* is the result of a humid climate, while *B* is the result of a dry climate.

(2) *A* is at a high elevation, while *B* is located at sea level.

(3) *A* is a plateau region, while *B* is a mountainous region.

(4) *A* is composed of igneous bedrock, while *B* is composed of sedimentary bedrock.   27 _____

28. The block diagram below shows a region that has undergone faulting.

Which map shows the stream drainage pattern that would most likely develop on the surface of this region?

     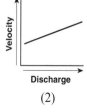

(1)   (2)   (3)   (4)   28 _____

29. Which graph best represents the relationship between the discharge of a stream and the velocity of stream flow?

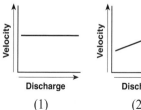   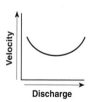   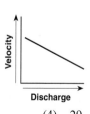

(1)   (2)   (3)   (4)   29 _____

30. The accompanying map shows the bend of a large meandering stream. The arrows show the direction of stream flow. Letters *A*, *B*, and *C* are positions on the streambed where erosion and deposition data were collected.

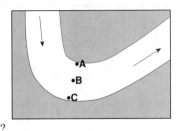

Which table best represents the locations where erosion and deposition are dominant and where an equilibrium exists between the two processes?
[A check mark (√) represents the dominant process for each lettered location.]

	Erosion	Equilibrium	Deposition
A		√	
B			√
C	√		

(1)

	Erosion	Equilibrium	Deposition
A	√		
B		√	
C			√

(3)

	Erosion	Equilibrium	Deposition
A			√
B	√		
C		√	

(2)

	Erosion	Equilibrium	Deposition
A			√
B		√	
C	√		

(4)

30 _____

31. The diagrams below represent four different examples of one process that transports sediments.

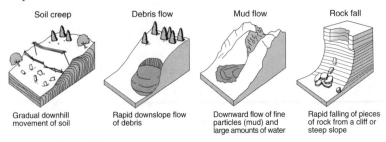

Soil creep — Gradual downhill movement of soil

Debris flow — Rapid downslope flow of debris

Mud flow — Downward flow of fine particles (mud) and large amounts of water

Rock fall — Rapid falling of pieces of rock from a cliff or steep slope

Which process is shown in these diagrams?
(1) chemical weathering   (3) mass movement
(2) wind action   (4) rock abrasion

31 _____

32. The accompanying cross section shows a stream flowing downhill. Points *A* through *D* are locations in the stream. At which point would most deposition occur?
(1) *A*   (2) *B*   (3) *C*   (4) *D*

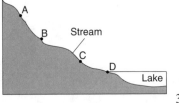

32 _____

33. A stream flowing at a velocity of 250 centimeters per second is transporting sediment particles ranging in size from clay to cobbles. Which transported particles will be deposited by the stream if its velocity decreases to 100 centimeters per second?
(1) cobbles, only
(2) cobbles and some pebbles, only
(3) cobbles, pebbles, and some sand, only
(4) cobbles, pebbles, sand, silt, and clay   33 _____

34. Which rock is sedimentary in origin and formed as a result of chemical processes?
(1) granite   (2) shale   (3) breccia   (4) dolostone   34 _____

35. The accompanying photograph shows an igneous rock. What is the origin and rate of formation of this rock?
(1) plutonic with slow cooling
(2) plutonic with rapid cooling
(3) volcanic with slow cooling
(4) volcanic with rapid cooling

35 _____

Part B–1
Answer all questions in this part.
Directions (36–50): For *each* statement or question, write in the space provided the *number* of the word or expression that, of those given, best completes the statement or answers the question. Some questions may require the use of the *Earth Science Reference Tables.*

Base your answers to questions 36 through 39 on the passage and diagram below. The diagram shows the orbits of the four inner planets and the asteroid Hermes around the Sun. Point *A* represents a position along Hermes' orbit.

The Curious Tale of Asteroid Hermes
It's dogma [accepted belief] now: an asteroid hit Earth 65 million years ago and wiped out the dinosaurs. But in 1980 when scientists Walter and Luis Alvarez first suggested the idea to a gathering at the American Association for Advancement of Sciences, their listeners were skeptical. Asteroids hitting Earth? Wiping out species? It seemed incredible.

At that very moment, unknown to the audience, an asteroid named Hermes halfway between Mars and Jupiter was beginning a long plunge toward our planet. Six months later it would pass 300,000 miles from Earth's orbit, only a little more than the distance to the Moon....

Hermes approaches Earth's orbit twice every 777 days. Usually our planet is far away when the orbit crossing happens, but in 1937, 1942, 1954, 1974 and 1986, Hermes came harrowingly [dangerously] close to Earth itself. We know about most of these encounters only because Lowell Observatory astronomer Brian Skiff rediscovered Hermes on Oct. 15, 2003.

Astronomers around the world have been tracking it carefully ever since....

Excerpted from "The Curious Tale of Asteroid Hermes"
Dr. Tony Phillips, *Science @ NASA*, November 3, 2003

36. When Hermes is located at position *A* and Earth is in the position shown in the diagram, the asteroid can be viewed from Earth at each of the following times *except*

(1) sunrise
(2) sunset
(3) 12 noon
(4) 12 midnight   36_____

Orbit of Asteroid Hermes

Mercury, Sun, Mars, Earth, Venus, A

(Not drawn to scale)

37. How does the period of revolution of Hermes compare to the period of revolution of the planets shown in the diagram?
(1) Hermes has a longer period of revolution than Mercury, but a shorter period of revolution than Venus, Earth, and Mars.
(2) Hermes has a shorter period of revolution than Mercury, but a longer period of revolution than Venus, Earth, and Mars.
(3) Hermes has a longer period of revolution than all of the planets shown.
(4) Hermes has a shorter period of revolution than all of the planets shown.   37 _____

38. Why is evidence of asteroids striking Earth so difficult to find?
(1) Asteroids are made mostly of frozen water and gases and are vaporized on impact.
(2) Asteroids are not large enough to leave impact craters.
(3) Asteroids do not travel fast enough to create impact craters.
(4) Weathering, erosion, and deposition on Earth have destroyed or buried most impact craters.   38 _____

39. According to the diagram, as Hermes and the planets revolve around the Sun, Hermes appears to be a threat to collide with
(1) Earth, only   (3) Venus, Earth, and Mars, only
(2) Earth and Mars, only   (4) Mercury, Venus, Earth, and Mars   39 _____

40. The accompanying map shows the location of Grenville-age bedrock found in the northeastern United States. In which New York State landscapes is Grenville-age bedrock exposed at Earth's surface?
(1) Erie-Ontario Lowlands and St. Lawrence Lowlands
(2) Catskills and Allegheny Plateau
(3) Tug Hill Plateau and Atlantic Coastal Plain
(4) Hudson Highlands and Adirondack Mountains

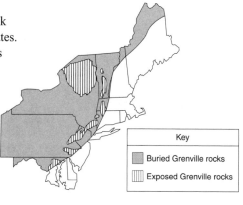

Key

▓ Buried Grenville rocks

▦ Exposed Grenville rocks

40 _____

Base your answers to questions 41 through 44 on the climate graphs below, which show average monthly precipitation and temperatures at four cities, A, B, C, and D.

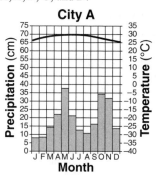

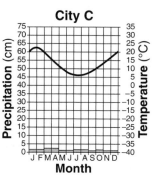

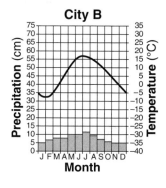

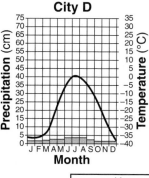

Key

▓ Precipitation

— Temperature

41. City *A* has very little variation in temperature during the year because city *A* is located
(1) on the dry side of a mountain (3) near the center of a large landmass
(2) on the wet side of a mountain (4) near the equator   41 _____

42. During which season does city *B* usually experience the month with the highest average precipitation?
(1) spring   (2) summer   (3) fall   (4) winter   42 _____

43. It can be concluded that city *C* is located in the Southern Hemisphere because city *C* has
(1) small amounts of precipitation throughout the year
(2) large amounts of precipitation throughout the year
(3) its warmest temperatures in January and February
(4) its warmest temperatures in July and August   43 _____

44. Very little water will infiltrate the soil around city *D* because the region usually has
(1) a frozen surface   (3) a small amount of runoff
(2) nearly flat surfaces   (4) permeable soil   44 _____

Base your answers to questions 45 through 47 on the block diagram below, which shows a portion of Earth's crust. Letters *A*, *B*, *C*, and *D* indicate sedimentary layers.

Key
Igneous rock
Contact metamorphism

45. Which event occurred most recently?
(1) formation of layer *A*
(2) formation of layer *D*
(3) tilting of all four sedimentary rock layers
(4) erosion of the igneous rock exposed at the surface   45 _____

46. The igneous rock is mostly composed of potassium feldspar and quartz crystals that have an average grain size of 3 millimeters. The igneous rock is most likely
(1) granite   (2) pegmatite   (3) gabbro   (4) pumice   46 _____

47. Which processes produced rock layer *B*?
(1) subduction and melting   (3) heat and pressure
(2) uplift and solidification   (4) compaction and cementation   47 _____

Base your answers to questions 48 through 50 on the map of Long Island, New York. *AB*, *CD*, *EF*, and *GH* are reference lines on the map.

Map

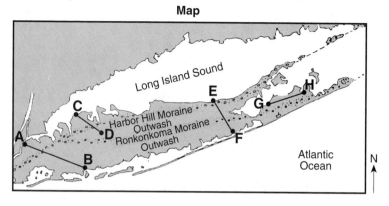

48. Which agent of erosion transported the sediments that formed the moraines shown on the map?

(1) water   (2) wind   (3) ice   (4) mass movement   48 _____

49. The cross section below represents the sediments beneath the land surface along one of the reference lines shown on the map.

Along which reference line was the cross section taken?

(1) *AB*   (2) *CD*   (3) *EF*   (4) *GH*   49 _____

50. A major difference between sediments in the outwash and sediments in the moraines is that the sediments deposited in the outwash are

(1) larger   (2) sorted   (3) more angular   (4) older   50 _____

Part B-2
Answer all questions in this part.
***Directions* (51-65): Record your answers in the spaces provided. Some questions may require the use of the *Earth Science Reference Tables*.**

Base your answers to questions 51 through 53 on the cross section on the next page, which shows limestone bedrock with caves.

51. In the empty box on the left side of the cross section below, draw a horizontal line to indicate the level of the water table. [1]

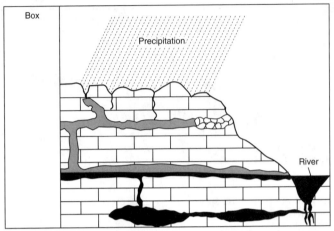

52. The precipitation in this area is becoming more acidic. Explain why acid rain weathers limestone bedrock, [l]

_____

_____

53. Identify *one* source of pollution caused by human activity that contributes to the precipitation becoming more acidic. [1]

_____

Base your answers to questions 54 and 55 on the accompanying data table, which provides information about four of Jupiter's moons.

Data Table

Moons of Jupiter	Density (g/cm³)	Diameter (km)	Distance from Jupiter (km)
Io	3.5	3630	421,600
Europa	3.0	3138	670,900
Ganymede	1.9	5262	1,070,000
Callisto	1.9	4800	1,883,000

54. Identify the planet in our solar system that is closest in diameter to Callisto. [1] _____

55. In 1610, Galileo was the first person to observe, with the aid of a telescope, these four moons orbiting Jupiter. Explain why Galileo's observation of this motion did *not* support the geocentric model of our solar system. [1]

_____

_____

Base your answers to questions 56 through 60 on the satellite image shown.

The satellite image shows a low-pressure system over a portion of the United States. Air-mass symbols and frontal boundaries have been added. Line *XY* is one frontal boundary. Points *A*, *B*, *C*, and *D* represent surface locations. White areas represent clouds.

56. Draw the proper symbol to represent the most probable front on line *XY*. [1]

57. State *one* process that causes clouds to form in the moist air along the cold front, [1]

_____

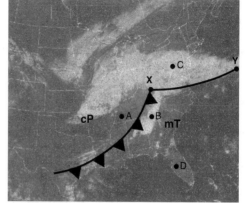

58. Describe *one* piece of evidence shown on the map that suggests location *A* has a lower relative humidity than location *B*. [1]

_____

_____

59. Explain why location *C* most likely has a cooler temperature than location *D*. [1] _____

_____

60. State the compass direction that the center of this low-pressure system will move over the next few days if it follows a normal storm track. [1]

_____

Base your answers to questions 61 and 62 on the accompanying cross section, which shows a portion of Earth's interior layers and the location of an earthquake epicenter. Letter *A* represents a seismic station on Earth's surface. Letter *B* represents a location in Earth's interior.

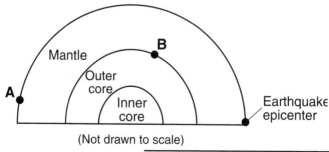

(Not drawn to scale)

June 2008

61. Explain why seismic station *A* receives *P*-waves but *not* *S*-waves from this earthquake. [1]

_____

_____

62. What is the approximate depth at location *B*? [1] _____

Base your answers to questions 63 through 65 on the accompanying diagram, which shows Earth's water cycle. Numbers indicate the estimated volume of water, in millions of cubic kilometers, stored at any one time in the atmosphere, the oceans, and on the continents. The yearly amount of water that moves in and out of each of these three portions of Earth is also indicated in millions of cubic kilometers.

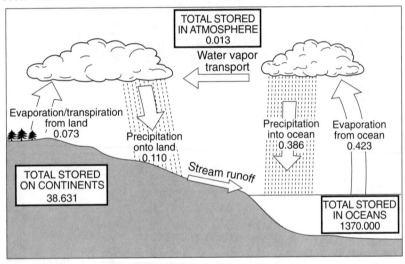

63. Calculate the total amount of water stored in the atmosphere, the oceans, and on the continents together at any one time. [1]

_____ millions of cubic kilometers

64. Explain why the yearly total precipitation over the oceans is greater than the yearly total precipitation over the continents. [1]

_____

_____

65. Describe *two* surface characteristics that will affect the rate of stream runoff into the ocean. [1]

Characteristic 1: _____

Characteristic 2: _____

Part C

Answer all questions in this part.

Directions **(66-84): Record your answers in the spaces provided. Some questions may require the use of the *Earth Science Reference Tables*.**

Base your answers to questions 66 through 68 on the accompanying data table, which shows the radioactive decay of carbon-14. The number of years required to complete four half-lives has been left blank.

Radioactive Decay of Carbon-14

Number of Half-Lives	Percentage of Original Carbon-14 Remaining	Time (years)
0	100	0
1	50	5700
2	25	11,400
3	12.5	17,100
4	6.3	
5	3.1	28,500
6	1.6	34,200

66. On the grid below, construct a graph that shows the radioactive decay of carbon-14 by plotting an **X** to show the percentage of original carbon-14 remaining after *each* half-life. Connect the **X**s with a smooth, curved line. [1]

67. How long does it take for radioactive carbon-14 to complete four half-lives? [1]

_____yr

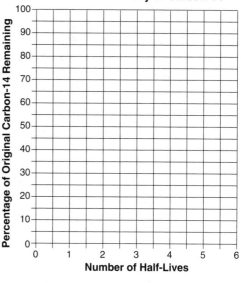

Radioactive Decay of Carbon-14

Percentage of Original Carbon-14 Remaining (y-axis, 0 to 100)

Number of Half-Lives (x-axis, 0 to 6)

68. The cross section below shows part of Earth's crust. The objects in parentheses indicate materials found within each rock unit or deposit.

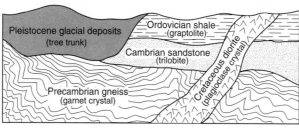

Pleistocene glacial deposits (tree trunk)
Ordovician shale (graptolite)
Cambrian sandstone (trilobite)
Cretaceous diorite (plagioclase crystal)
Precambrian gneiss (garnet crystal)

Which object in parentheses could be accurately dated using carbon-14? Explain your answer. [1]

Object: _____

Explanation: _____

Base your answers to questions 69 through 73 on the passage and cross section below, which explain how some precious gemstones form. The cross section shows a portion of the ancient Tethys Sea, once located between the Indian-Australian Plate and the Eurasian Plate.

Precious Gemstones

Some precious gemstones are a form of the mineral corundum, which has a hardness of 9. Corundum is a rare mineral made up of closely packed aluminum and oxygen atoms, and its formula is Al_2O_3. If small amounts of chromium replace some of the aluminum atoms in corundum, a bright-red gemstone called a ruby is produced. If traces of titanium and iron replace some aluminum atoms, deep-blue sapphires can be produced.

Most of the world's ruby deposits are found in metamorphic rock that is located along the southern slope of the Himalayas, where plate tectonics played a part in ruby formation. Around 50 million years ago, the Tethys Sea was located between what is now India and Eurasia. Much of the Tethys Sea bottom was composed of limestone that contained the elements needed to make these precious gemstones. The Tethys Sea closed up as the Indian-Australian Plate pushed under the Eurasian Plate, creating the Himalayan Mountains. The limestone rock lining the seafloor underwent metamorphism as it was pushed deep into Earth by the Indian-Australian Plate. For the next 40 to 45 million years, as the Himalayas rose, rubies, sapphires, and other gemstones continued to form.

69. Which element replaces some of the aluminum atoms, causing the bright-red color of a ruby? [1]

_____

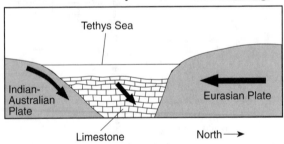

A Portion of the Tethys Sea 50 Million Years Ago

Tethys Sea

Indian-Australian Plate

Eurasian Plate

Limestone

North ⟶

70. State *one* physical property of rubies, other than a bright-red color, that makes them useful as gemstones in jewelry. [1]

_____

71. Identify the metamorphic rock in which the rubies and sapphires that formed along the Himalayas are usually found. [1] _____

72. During which geologic epoch did the events shown in the cross section of the Tethys Sea occur? [1] _____ Epoch

73. What type of tectonic plate boundary is shown in the cross section? [1] _____

Base your answers to questions 74 through 77 on the diagrams below. Diagram 1 shows Earth's location in its orbit on the first day of each of the four seasons, labeled *A* through *D*. Diagram 2 shows a north polar view of Earth on March 21. Point *E* represents a location on Earth's surface. Longitude lines are shown at 15° intervals.

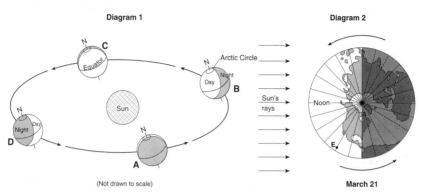

Diagram 1

Diagram 2

(Not drawn to scale)

March 21

74. How does the altitude of the Sun at solar noon appear to change each day for an observer in New York State as Earth moves from position *A* to position *B* to position *C*? [1]

_____

_____

_____

75. Explain why the duration of insolation is 12 hours at both the Arctic Circle and the equator when Earth is at position *C*. [1]

_____

_____

_____

76. Describe *one* piece of evidence shown in the diagram which indicates that the Northern Hemisphere is experiencing winter at position *D*. [1]

_____

_____

_____

77. State the hour of the day at point *E*. [1] _____

Base your answers to questions 78 through 80 on the map below. The map shows the water depth, measured in feet, at the north end of one of the Finger Lakes. Points *A* and *B* are locations at the lake's shoreline. Points *X* and *Y* are locations on the bottom of the lake.

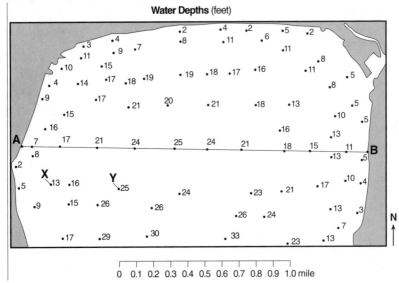

Water Depths (feet)

0 0.1 0.2 0.3 0.4 0.5 0.6 0.7 0.8 0.9 1.0 mile

78. On the map above, draw the 20-foot-depth isoline. The isoline must extend to the edge of the map. [1]

79. On the grid below, construct a profile along the line from point *A* to point *B*. Plot the depth along line *AB* by marking an **X** at each numbered point where a water depth is shown. Complete the profile by connecting the **X**s with a smooth, curved line. The **X**s for point *A* and point *B* have been plotted. [2]

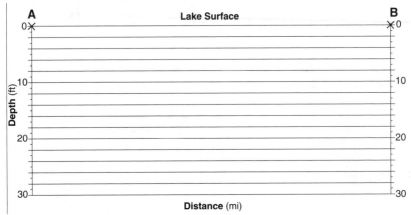

Lake Surface

Depth (ft)

Distance (mi)

80. Calculate the gradient between point *X* and point *Y*.
Label your answer with the correct units. [1]   Gradient:_____

Base your answers to questions 81 through 84 on the map below, which shows a portion of southwestern United States. On January 17, 1994, an earthquake occurred with an epicenter at Northridge, California.

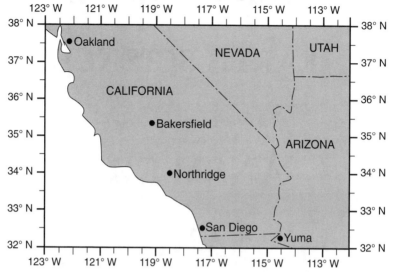

81. State the latitude and longitude of Northridge, California. Include the correct units and compass directions in your answer. [1]

Latitude: _____   Longitude: _____

82. Explain why earthquakes are common in this region of California. [1]

_____

_____

_____

83. Of the cities shown on the map, explain why Oakland was the last city to receive *P*-waves from this earthquake. [1]

_____

_____

_____

84. List *two* actions that a homeowner could take to prepare the home or family for the next earthquake. [1]

Action 1: _____

_____

Action 2: _____

_____

PHYSICAL SETTING

EARTH SCIENCE

ANSWERS

AND

EXPLANATIONS

An Earth Science Reference Table (RT) is quoted throughout this section. The Earth Science Reference Table can be found in the back of this booklet. Other Earth Science Reference Tables may be used in its place.

1. 2   Open the RT to the Solar System Data chart. Use the Equatorial Diameter column to arrange the given objects from the smallest to largest by diameter. Using this chart, Pluto has the smallest diameter, while Mars has the largest diameter of the given choices.

2. 3   Seasons are caused by the 23½° tilt of the Earth's axis and revolution. Because of this tilt, direct rays move north reaching the Tropic of Cancer (23½°N) on June 21. An increase of tilt of the Earth's axis to 35° would cause direct rays to move further north, to the 35°N latitude line.These direct (stronger in intensity) rays being closer to NYS would cause higher summer time temperatures for us. During the winter direct rays move to the Tropic of Capricorn (23½°S) on December 21. Increasing the tilt would cause the direct rays to move further south, to the 35°S latitude line. Being further from NYS, this would cause colder temperatures.

3. 4   Open to the Luminosity and Temperature of Stars chart in the RT. On the Luminosity axis, the brightest stars are near the top, with the dimmer stars near the bottom. Locate our Sun. From the given choices, only Aldebaran has a higher luminosity, and a lower temperature, around 4,000°C, than our Sun.

4. 1   According to the Big Bang theory, our universe is still expanding and galaxies are still moving away from each other at tremendous speeds. Evidence of this motion is obtained by studying the red shift of a distant galaxy's spectral lines. As galaxies move away from our galaxy the wavelength of these spectral lines will become longer and move towards the red end of the spectrum.

5. 2   The higher the noon Sun is, the more intense, or stronger the Sun's radiation will be. On the first day of summer, June 21, direct rays are furthest north of the equator hitting the Tropic of Cancer (23½°N). On this day at noon, the Sun's altitude is the highest for NYS, resulting in the greatest intensity for anywhere in NYS. Remember, even though the Sun reaches its highest altitude at this time, it is never directly overhead for any location in NYS.

6. 3   The equation for eccentricity is found in the Equation section of the RT ($e = d/\ell$). Using the given information, ℓ = 7.5cm and d = 5.5cm, substitute into the equation. This gives e = 5.5cm/7.5cm or 0.7. Remember, eccentricity does not have a unit, it is a number only. This procedure is used when the elliptical orbit is to scale as the Regents question will be.

7. 1   Throughout the year, the Sun's direct rays move between the Tropic of Cancer (23½°N) and the Tropic of Capricorn (23½°S). All other areas receive indirect sunlight. The Earth's polar regions, 90°N and 90°S, being the furthest from the equator, receive the least direct rays (low-angle insolation) resulting in very cold temperatures. Answer 2 is wrong since the polar regions during the summer experience 6 months of 24 hours of sunlight.

8. 3   Oceans modify the climate of coastal cities. During the summer, the ocean will generally be cooler than the land producing a cooling ocean breeze. During the winter, the ocean will generally be warmer than the land. This additional heat produces slightly warmer winter temperatures. Inland cities, at the same latitude and elevation, do not have the oceans to modify their temperatures, thus they tend to have hotter summers and colder winters than a coastal city.

9. 4   A swinging pendulum will appear to change direction over a period of time (usually within hours). This apparent shift in direction is caused by the Earth rotating under it. This is a proof of Earth's rotation – the Foucault pendulum.

10. 1   At solar noon for NYS, the Sun is at its highest position in the sky and the Sun is located in the southern section of the sky. This is so, since the Sun's direct rays are to the south of NYS, traveling between the Tropic of Cancer (23½°N) and the Tropic of Capricorn (23½°S). Since the Sun is in the southern part of the sky at noon, shadows will point north. Remember, shadows are always opposite the direction of the Sun

11. 4   The position of the moon next to where it says "Sun's rays" is the new moon which can't be seen since it is up during the day. As the moon revolves around the Earth from this position, as shown by the arrows, the amount of moon light we see from Earth slowly increases until the moon is full, which will be opposite the new moon position. Position 1, is called the new or waxing gibbous moon and appears just before the full moon and would look like diagram 4.

12. 4   During the day, the Earth absorbs radiation from the Sun mostly in the visible and ultraviolet range. At night, the Earth will radiate off longer wavelengths back to space. This longer wavelength radiation is mostly infrared. Open to the Electromagnetic Spectrum chart in the RT and here it shows the different types of electromagnetic radiation and their respective wavelengths. Notice that infrared has a longer wavelength compared to ultraviolet or visible radiation.

13. 1   Plains and plateaus have an underlying bedrock of horizontal sedimentary layers. Plateaus have higher elevations causing them to be dissected, or cut up, by agents of erosion (like water) causing steep slopes. This type of landscape is widespread in NYS in areas like the Allegheny Plateau as shown in the Generalized Landscape Region of NYS chart in the RT. Mountain ranges often have tilted or folded bedrock structure.

14. 1   Winds in a low-pressure system move in a counterclockwise motion, inward toward the center, while winds in a high-pressure system blow in a clockwise direction, out from the center. Diagrams 1 and 2 show counter-clockwise direction, but only diagram 1 shows the winds moving inward. Also, diagram 1 shows the correct millibar readings; they must decrease toward the center of a low.

15. 4   As the air and the dewpoint temperatures move closer together, the relative humidity increases. When these temperatures are the same, the RH will be 100% and since the dewpoint has been reached, condensation will begin, producing clouds and increasing the chance of precipitation.

16. 2   Open to the Surface Ocean Currents chart in the RT. The given key shows warm currents having a darkened arrow. Notice that the Brazil Current, located on the eastern side of South America, and the Agulhas Current, which flows along the eastern section of Africa, are both warm currents.

17. 3   Infiltration is the passing of water through Earth's surface material. This will occur if the ground is permeable, having open air spaces or pores. Infiltration increases if the ground is not very steep allowing the water time to enter the pores.

18. 1   A delta is produced when sediments are deposited when a stream enters a large body of water. Eventually, over many years, new land is made. In a deposition system, the larger sediments are released or dropped first, than the finer ones (silt and clay) are deposited further out in the water. Graph 1 shows this relationship.

19. 4   Open to the Igneous Rock chart in the RT and locate pumice in the upper left section of this chart. Directly to the left of pumice is given the environment that produced it - Extrusive (volcanic) environment. Pumice is produced when lava is ejected from an erupting volcano, and cooling very rapidly trapping air.

20. 3   Open to the Metamorphic Rock chart in the RT. On the left side of this chart locate Foliated Texture then Banding. Follow this row to the Rock Name and gneiss is given. From the Composition Section, the minerals that are found in gneiss are given, which include amphibole, quartz and feldspar.

21. 4   Open to the Igneous Rock chart in the RT and locate Pegmatite. Move to the right and in the Texture section it states that a very coarse texture has crystals (grain sizes) larger than 10 mm. Using the ruler located on the front page of the RT, the crystals in this picture are very coarse being larger than 10 mm. The given texture for pegmatite is very coarse. To grow such large crystals, magma must cool and solidify very slowly deep underground.

22. 2   Open to the Igneous Rock chart in the RT and locate granite. The minerals found in granite are found directly under granite in the Mineral Composition chart. The given minerals in the question match the minerals found in granite.

23. 4   The photograph shows the colorless mineral pieces have at least one smooth surface plane. This is the mineral property of cleavage. Fracture is the opposite; a mineral exhibiting uneven breakage planes.

24. 4   Open the RT to the Properties of Common Minerals and find quartz, halite and calcite. These three minerals may be colorless as mentioned in the Common Color column. Quartz can be eliminated, since the shown crystal shape of quartz does not match the photograph. Comparing the given diagram, in the RT, of calcite and halite, halite exhibits cubic cleavage, which matches the photograph.

25. 1   Coal is believed to be formed when a tropical forest that dies off, gets covered by sediments, slowly being compressed by the weight of sedimentary layers (see coal in the Sedimentary Rock chart). Much of this process occurred during the Carboniferous period. Open up to the Geologic History of NYS chart in the RT and locate the Carboniferous Period, which the Mississippian is a part of. Move to the right side of this chart to the Inferred Position of Earth's Landmasses, and notice the position of North America on the Devonian/Mississippian diagram. Pennsylvania appears to be on the equator, the environment needed to grow tropical forests. Since that time, North America has drifted northwestward as shown by the global diagrams in this column due to tectonic plate movements.

26. 3   Open to the Inferred Properties of Earth's Interior in the RT. Locate the boundary line between the Stiffer Mantle and the Outer Core. Using this location, follow the dashed line down into the Temperature graph. This dashed line intersects the Actual Temperature line at the 5000°C position.

27. 1   Open to the Electromagnetic Spectrum chart in the RT. The visible section of the chart is expanded into the different colors based on wavelengths. The shortest wavelengths are located to the left, with violet having the shortest wavelength of the visible spectrum.

28. 2 The diagram is showing the collision of two convergent plates in which the denser Oceanic crust is subducting under the less dense Continental crust. Open to the Tectonic Plates chart in the RT. Using the given key for Convergent Plate Boundary, this boundary exists between the Nazca Plate and South American Plate region.

29. 3 Open to the Physical Constants chart on the first page of the RT. In the Properties of Water chart, it shows that when water freezes, 80 calories/gram are released or lost by the water.

30. 1 Open to the Generalized Bedrock Geology of NYS chart in the RT. Locate the key (bottom left) and the last two key boxes match the bedrock shown at the western shores of Lake Champlain (found in the NE section of NYS). Directly to the right of these key boxes is given the type of rock for these key boxes – Intensely Metamorphic Rocks.

31. 4 In time, all radioactive elements decay into non-radioactive elements. Radioactive Carbon-14 atoms will change over time into non-radioactive N-14 atoms (see Radioactive Decay Data chart). Graph 4 correctly shows the decay curve of Carbon-14 over time as it gets weaker and weaker.

32. 1 Open up to the Geologic History of NYS chart in the RT and locate the Cenozoic Era. Moving to the right, the chart shows that the Cenozoic Era started around 65 million years ago. Now go to the Eon section and locate the time bar. The bottom of this bar represents the estimated time for the beginning of the Earth or 4600 millions of years ago. So, what % is 65 million years of 4600 million of years?

Solve by dividing: $\dfrac{65\,\text{mya}}{4500\,\text{mya}} = 0.014$

To change to percent multiply by 100.

Solving: $0.014 \times 100 = 1.4\%$

33. 2 An unconformity occurs when layers are eroded away producing a geologic time gap. This is usually shown by an uneven erosion line. Open to the Geologic History of NYS chart in the RT and find the Silurian period. Under the Silurian Period should be the Ordovician period. This whole geologic period is missing in the diagram, causing the unconformity. From the chart, the Ordovician period started 490 millions of years ago (see the time bar just to the right) and ended 443 millions of years ago. Subtracting these numbers, $490 - 443 = 47$ millions of years is at least missing. This would be the minimum time of this unconformity.

34. 4   Stream drainage patterns are caused by the topography or landscape of a region. Notice that these streams flow parallel with smaller tributaries flowing 90 degrees to the main stream. The main steams must be following valleys, which separate the streams and prevents them from joining. Eroded, folded mountains produce valleys with gaps, making this type of drainage pattern. Diagram 4 matches this description and given diagram.

35. 4   A mT air mass would be wet (maritime) and warm (tropical) and be given the initials mT. A maritime polar air would be moist (maritime) and cold (polar) and will be given the initials of mP. Warmer air can hold more moisture than colder air, thus an mP air mass would contain less moisture that a mT air mass.

Part B-1

36. 4   A barometer is an instrument that is sensitive to changes in the weight of the air, and records air pressure in millibars. A sling psychrometer is used to get the dewpoint. A wind vane shows the air direction, while a thermometer records the air temperature.

37. 2   Where isobars are close together high winds will be found. This is so, since closely spaced isobars indicate a strong pressure gradient within the atmosphere causing wind.

38. 3   Position *D* is between the 1008 mb and 1012 mb isobar lines making a reading of 1010 mb an accurate estimation of the air pressure at position *D*.

39. 1   Fossils are almost always preserved in sedimentary rocks. The jellyfish, being a marine animal would most likely be floating above sand that is normally present in lagoons. Open to the Rock charts in the RT and sandstone is listed as a sedimentary rock that would have preserved the dead jellyfish, while granite and pumice are igneous rocks and slate is found in the metamorphic rock chart. Metamorphic and igneous environments tend to destroy fossil evidence.

40. 3   In the reading, it states the jellyfish were deposited during the Cambrian period. Open to the Geologic History of NYS in the RT. Go down to the Cambrian period and follow this row over to the Time Distribution of Fossils column. Notice that the brachiopod and gastropod bars start in the Cambrian period, thus these species lived during this geologic time period along with jellyfish.

41. 4   Almost all fossils are found in sedimentary rocks. Preserved ripple marks, the ones you see being made by ocean waves, are a strong indicator that a tide or storm washed the jellyfish up and out of their surroundings. These ripple marks are associated with a water environment which is usually needed for the making of sedimentary rocks. Answer 2, is wrong since "a distorted crystal structure" is a clue of metamorphism where heat/pressure distorts existing crystals.

42. 4   The topographic map shows contour lines with a contour interval of 100 feet. Point **X** is above the 1600 contour line and must be lower than the 1700 foot marker since no 1700 foot contour line exists. This gives us two choices: answers 3 and 4. The highest elevation could be 1 foot less than 1700 feet, or 1699 feet high.

43. 2   Open to the Equation section in the RT and locate the Gradient equation. The change in field is the difference of the elevation between C and D. To get the distance, measure the miles from C to D using the given scale. Substituting into the equation gives:

$$\text{Gradient} = \frac{1500' - 1000' \ \text{(change in field)}}{2 \ \text{miles} \qquad \text{(distance)}} \Rightarrow \frac{500'}{2 \ \text{miles}} \Rightarrow 250 \, \text{ft} / \text{mile}$$

44. 1   The profile is the side view of a region. Starting at position A, the elevation is 900 feet. The profile line crosses a section of the lake between the two 900 contour lines. This would indicate a slight depression is needed to contain the water as shown by profile 1. From the second 900 foot contour line, the elevation rises to over 1600 feet just to the left of point **X**. On the backside of the mountain, the contour lines are closer together indicating a steeper gradient, or steeper slope. The elevation drops quickly to 900 feet where position B is located. Again, profile 1 matches these elevation readings. Notice profile 2 does not accommodate a slight depression for the northwestern section of the lake.

45. 2   Open to the Generalized Landscape Regions of NYS in the RT. Match up the Susquehanna – Chesapeake and Delaware watersheds with the landscape regions of NYS. They are found on the Allegheny Plateau of which the Catskills is a part.

46. 1   Open to the Generalized Bedrock Geology of NYS in the RT. Locate the Genesee River, in the western part of NYS. From the given watershed map, the Genesee River is being drained by the Ontario–St Lawrence watershed.

47. 2   Open to the Generalized Bedrock Geology of NYS in the RT. Matching up the area of the Ontario-St Lawrence watershed to the given key area of this chart. It shows that the southern area of this watershed was formed during the Devonian period, while most of the northern sections of this watershed were formed during the older Silurian and Ordovician periods.

48. 3   When the northern axis is tilted at the Sun, the Northern Hemisphere will be experiencing summer; this is position *B*. Six months later, the northern axis would be tilted away from the Sun producing indirect rays for the Northern Hemisphere, causing the winter season. This is position *D*. This would make position *C*, fall, and position *A*, spring – the equinoxes.

49. 1   The Sun takes 365.26 days to complete one revolution (see the Solar System Data chart in the RT). One complete orbit, or revolution, is equal to 360 degrees. This works out to be very close to 1 degree of revolution for each day. Answer 3, 15° per hour is the speed of the Earth's rotation.

50. 4   Throughout the year, new constellations are seen at different times of the year. These are called seasonal constellations. Orion, "The Hunter" is a famous winter constellation. Six months later, the Earth revolves to the opposite position of our orbit, (the summer position) causing Orion to be up during the day making it invisible being outshined by our Sun. But at this time and position in our orbit, new constellations will be seen. Answers 1 and 2 are proofs of rotation, which can be observed in hours.

Part B-2

51. Acceptable answers include, but not limited to:
    Warm, moist air is less dense than cold, dry air.
    or Cold air is moving under the warm air and forcing the warm air upward.

    Explanation: The cP, continental polar, air mass contains dry (continental) cold (polar) air which is colliding with the mT, maritime tropical, air mass which contains moist (maritime), warm (tropical) air. At the frontal boundary, shown by the dark curving line, warm moist air is rising. This occurs, since warm air is less dense and is forced upward by the colder more dense air.

52. Acceptable answers include, but not limited to:
    The air is expanding.
    or The rising air cools to the dewpoint.
    or Condensation has started.

    Explanation: The diagram shows the warm, moist air (mT) rising as more dense colder air (cP) forces it upward. As air rises, it expands. This produces fewer collisions of air molecules causing the temperature to drop. As the temperature drops, it eventually will reach the dewpoint temperature causing condensation (G→L). These small liquid drops are the making of a cloud.

53. Acceptable answers include, but not limited to:
Gulf of Mexico *or* a warm ocean surface

Explanation: A mT air mass is a maritime tropical air mass. Maritime (m) indicates the air mass contains much moisture, while tropical (T) indicates a warm temperature. Both of these conditions exist around the source area of the Gulf of Mexico.

54. Answer: 44.5°N (latitude) and 73.7°W (longitude).

Explanation: The latitude reading is given first, and is the reading north (N) of the equator. Position **X** is between 44.4°N and 44.6°N. Thus, position **X** latitude is very close to 44.5°N. The longitude for position **X** is the reading west (W) of the Prime Meridian. Position **X** longitude is between 73.6°W and 73.8°W. Thus, position **X** longitude is very close to 73.7°W.

55. Answer: 3

Explanation: To locate the epicenter of an earthquake, three seismographs are needed. Using the difference in arrival time of the *P* and *S* waves, the distance to the epicenter from each seismograph can be found. Using the distances from the 3 seismographs and drawing 3 circles on a map, representing the distances obtained, there will be an intersection point of all 3 circles. This is the location of the epicenter. Remember, one seismograph will give you distance, but not direction to the epicenter.

56. Answer: Peru is closer to the epicenter.

Explanation: An earthquake releases seismic waves which move in all directions away from the epicenter. As they travel they lose energy, becoming weaker as distance increases. Peru being closer to the epicenter will normally feel more intense shaking than farther cities, like Lake Placid.

57. Answer: 3 min 0 sec *or* 3:00± 20 second

Explanation: Open to the Earthquake chart in the RT. Go to the 1800 km line and move up this line until it intersects the *P*-wave line. It intersects at 3:40 (3 minutes 40 seconds). Continue upward on the 1800 km line until it intersects the *S*-wave line. It intersects at 6:40. The difference of these two arrival times is 3 minutes (6:40- 3:40 = 3 minutes.)

58. Answer: Rock unit VII: 3   Rock unit VIII: 1   Rock unit IX: 2

Explanation: In relative dating, normally the older rocks are at the bottom, and younger rocks are above them and intrusions are younger than the rocks they cut across. This makes rock unit VIII containing the trilobite the oldest, being on the bottom. Later on in geologic history magma intruded producing rock unit IX. This makes the intrusion (IX) younger than rock unit VIII. Since rock unit VII is over rock unit VIII and IX it must be younger than both.

59. Answer: *Cryptolithus* and *Phacops*

Explanation: Open to the Geologic History of NYS chart in the RT. The fossil shown in rock unit VIII, is an Elliptocephala (letter A) shown on the top left of this chart. Near the center of this chart is the Time Distribution of Fossils section. Find the bar with letter A on it. This is the trilobite bar and shows that species A lived in the Cambrian period. Moving up on this trilobite bar are the index fossils shown by letter B and C. Match the name of these trilobite index fossils from the diagrams given at the left top of this chart.

60. Acceptable answers include, but not limited to: marine *or* ocean *or* water

Explanation: The given fossils in the shale and limestone layers represent animals that lived in shallow ocean seas. This is referred to as a marine environment. The fossil in rock unit I, a plant, would have lived in a land or terrestrial environments.

61. Acceptable answers include, but not limited to:
The type 3 stream meanders more.
or The type 3 stream occupies a wider floodplain.
or The type 1 stream has a straighter course.
or The type 1 stream shows more down-cutting, while
    the type 3 stream shows more meandering.

Explanation: The faster the water flows, the more direct its path will be, causing a straighter stream channel (type 1 channel). These straighter channels, undergo much down-cutting, are present in regions with steep slopes or high gradient, such as mountainous areas. When the gradient is less, water velocity slows down resulting in the water having less energy. Now the stream starts to erode side to side causing a meandering stream channel (type 3 channel) and possibly producing a broad floodplain.

62. Acceptable answers include, but not limited to:
Stream velocity is greater on the outside of the meandering channel.
or Stream flow is slower on the inside of the meandering channel.
or Water is moving faster on the outside of a meander curve.

Explanation: As water enters a curved channel, its speed increases on the outside of the curve, while slowing down on the inside of the curve. The faster water, on the outside curve, will have more energy causing more erosion, while the slower water will produce more deposition. These processes working together produce a meandering stream.

63. Acceptable answers include, but not limited to:
These tumbling cobbles and pebbles were abraded against
other transported rocks and the stream channel.
or Abrasion occurred as the rocks bounced and rolled
along the bottom of the streambed.
or Sharp corners and edges were knocked off, scraped, and/or worn down.
or Grinding against other sediment and rocks.

Explanation: Abrasion is the grinding and wearing away of rocks. Water is the
agent that carries rocks, but it is the collision with other sediments or with the
stream channel that chip away the sharp edges of the rocks. Eventually, by this
abrasive action the cobbles and pebbles become smoother, rounder and smaller.

64. Answer: 1002.1mb

Explanation: Open to the Weather Map Symbols chart in the RT. Locate the
barometric pressure in the upper right hand corner of the station model. This
reading 196 was originally 1019.6mb and then abbreviated by dropping the
decimal and the 10. From the station model in the exam booklet, the
abbreviated barometric reading is 021. To change this back to the original
barometric reading, we need to add a 10 and a decimal, this makes 1002.1mb.
Add a 10 and a decimal to any abbreviated reading that is less than 500, and
add a 9 and a decimal to any abbreviated reading that is more than 500.

65. Answer: south southeast (SSE) at 25 knots   *or*   southeast (SE) at 25 knots

Explanation: Open to the Weather Map Symbols chart in the RT. The wind
direction is always where the wind is coming from and is shown by the shaft
on a station model. The shaft (representing the wind direction) is read as if it
was entering the center of the station model. This makes the direction of the
wind from the southeast. Each whole feather represents 10 knots and a half
feather is 5 knots. This makes the wind speed at 25 knots.

Part C

66.

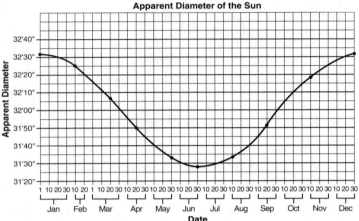

Explanation: Each line of the *y*-axis represents 05 seconds (05"). Locate the correct date along the *x*-axis, and then move upward to find the correct apparent diameter for that date along the *y*-axis. Accurately estimate the apparent diameter for *y* readings that fall between lines. Once all dots are positioned, connect with a smooth curve line. Do not use a ruler.

67. Acceptable answers include, but not limited to:
Earth has an elliptical orbit.
or The distance between the Sun and Earth varies in a cyclic manner.
or Earth is closest to the Sun during New York State's winter.
or The Sun is farthest from Earth during New York State's summer.

Explanation: Changes in the apparent diameter are caused by distance changes. The closer the object is, the larger its apparent diameter would be. The Earth's orbit around the Sun is elliptical. This causes the Earth to be slightly closer in the winter and slightly farther away from the Sun in the summer. When the Earth is closer, the Sun will appear to look slightly larger in winter, and smaller during summer.

68. Acceptable answers include, but not limited to:
water vapor (H_2O)   *or* methane (CH_4)
or carbon dioxide (CO_2)   *or* nitrous oxide (N_2O)
or ozone (O_3)   *or* chlorofluorocarbons (CFCs)

Explanation: Greenhouse gases are atmospheric gases that have the ability to absorb or trap Earth's infrared waves. Carbon dioxide is a greenhouse gas, released extensively by burning fossil fuels. Greenhouse gases are a major contribution of global warming. The other listed answers are also considered greenhouse gases.

69. Acceptable answers include, but not limited to:
    unsorted deposits   *or* moraines
    or drumlins   *or* till
    or mixed sediment sizes *or* glacial erratics/boulders
    or striated sediment

    Explanation: A glacier acts like a large bulldozer, carrying and dragging much sediment as it plows slowly forward. These transported sediments, will be unsorted in all different sizes and shapes, from the large boulders called erratics, to the smallest – clay particles. When the glacier melts or retreats, these sediments are deposited in an unsorted manner, and is given the term glacier till. These piles of sediments may produce large hills called moraines or smaller mounds called drumlins.

70. Acceptable answers include, but not limited to:
    a high elevation above sea level
    or mountains
    or a plateau

    Explanation: Glaciers exist in colder climates found in higher latitude areas. For a glacier to exist near the equator, the altitude must be very high to maintain the lower temperatures needed for the formation of a glacier. This would only occur in mountainous landscapes, where the elevation reaches above 15,000 feet, and temperatures are low year round.

71. Acceptable answers include, but not limited to:
    Stop burning fossil fuels.
    or Reduce the burning of tropical rain forests.
    or Reduce greenhouse-gas emissions.
    or Use more alternative energy sources such
        as solar collectors and wind turbines.
    or Do more car pooling.

    Explanation: Global warming has been linked to many causes. One of the biggest cause is the burning of fossil fuel: coal, gasoline, wood, oil, etc., that release large amounts of greenhouse gases (see explanation question 68). Any action that would reduce our dependence on fossil fuel would be a positive step in reducing the emission of greenhouse gases. In burning tropical forests, carbon dioxide is released - a greenhouse gas. On the other hand, a living tree removes carbon dioxide by respiration.

72. Answer: 450 ft ±50 ft

    Explanation: Each line on the y-axis represents 100 feet. The cross section chart shows that the bottom of the sediment (the blackened area) in Seneca Lake is just below the –600 foot line, around –625', while the top of the sediment is just above the –200 foot line, around –175'. This makes a difference of: $(-625) - (-175) = 450$' of sediment from the bottom to the top.

73. Acceptable answers include, but not limited to:
    The continental ice sheets generally moved from north to south.
    or glacial erosion
    or The original stream valleys had a north-south orientation.

    Explanation: During the ice age, a large sheet of ice, a glacier, slowly moved southward out of Canada into America. This north to south movement of the ice sheet, gouged out valleys, which later filled in with the glacier's melt water. Thus, the N to S orientation of the Finger Lakes is a result of the direction that the glaciers moved.

74. Acceptable answers include, but not limited to:
    Water has a higher specific heat than land.
    or Water heats and cools slowly because of its higher specific heat.
    or Bodies of water change temperature more slowly than surrounding land.

    Explanation: Open to the Specific Heats of Common Materials chart in your RT. Specific heat is how fast a material heats up or cools down compared to water. The higher the specific heat, the slower a substance heats up or cools down. Water has the highest specific heat (1.0) of the listed materials, meaning it heats up and cools down more slowly compared to the other substances like land with a specific heat around 0.2. During the winter, as the land cools down and begins to become frozen; water will cool down slower remaining a liquid. Thus, water remains unfrozen longer due to its high specific heat. (Note: for the specific heat for land, use granite's specific heat number, 0.19.)

75. Acceptable answers include, but not limited to:
    Sediments were deposited in water and compressed.
    or compaction and deposition
    or burial and cementation
    or chemical precipitation and evaporation.

    Explanation: Open to the Generalized Bedrock of NYS chart in the RT. The Finger Lakes are located on Devonian bedrock as shown by the given key area. In this key area to the right of Devonian, the rock type is given – Dominantly Sedimentary Origin. Now, open to the Rock Cycle chart in the RT. Following the arrow after the word "Sediments", the processes to produce a sedimentary rock are given: Deposition, Burial, Compacted and Cementation. From the Scheme for Sedimentary Rocks chart in the RT, evaporates and chemical precipitate are given here. They are acceptable processes to produce a sedimentary rock.

76-77.

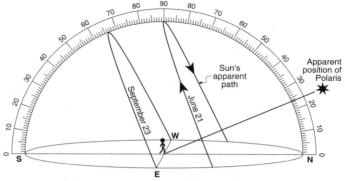

Sailor's Observations on the Deserted Island

76. Explanation: The Sun rises in the east and sets in the west. Draw an arrow on the June 21 path to show this direction.

77. Answer: The above diagram shows the height of the noon Sun to be 66.5°.

Explanation for the correct points on the horizons: On September 23, the Sun's direct rays (90°) follow the equator, which runs due (exactly) east and west, and making it the first day of autumnal equinox. On both of the equinoxes (March 21/Sept. 23), the Sun will rise due east and set due west.

Explanation for the correct altitude of the noon Sun: During the first day of summer, June 21, the Sun's direct rays are the furthest north traveling up to the Tropic of Cancer, 23.5°N latitude line. At noon the Sun would be directly overhead (90°) if you were on this latitude line. This is represented in the diagram for June 21. Each day after this, the Sun's direct rays slowly travel further south reaching the equator on the first day of fall, September 23. The Sun's direct rays have traveled from the 23.5°N line to the 0 degree line, the equator, for a total of 23.5° to the south. Subtracting: 90° – 23.5°, you have arrived at the height of the fall Sun, 66.5° for this position.

78. Acceptable answers include, but not limited to:
23°30' N   *or*   23.5°N   *or*   23½°N

Explanation: The angle of Polaris above the northern horizon is equal to the observer's latitude. From the diagram, the sailor measures Polaris' altitude as 23.5°. Using the above rule, this makes his latitude 23.5°N.

79. Acceptable answers include, but not limited to: 15   *or*   15°E   *or*   15°W

Explanation: There is a one hour separation of solar noon between his ship and the island. The Sun appears to travel 15 degrees for each hour as the Earth rotates, therefore every hour of separation must equal 15 degrees of longitude. This is how time zones are set-up.

80. Acceptable answers include, but not limited to:
Location *B* is at a transform fault, and location *C* is at a subduction boundary.
or Location *B* has horizontal plate movement, but location *C*
   has vertical plate movement.
or There is a transform plate boundary at *B*.
or There is a subduction plate at *C*.

Explanation: Open to the Tectonic Plates chart in the RT and review the "key" information for convergent plate boundaries. Location *C* is at the Peru-Chile Trench. Here the Nazca Ocean Plate and South American Continental Plate are converging. The oceanic plate will dive down (subduct) under the continental plate producing a trench. As subduction continues, the oceanic plate goes deeper into the crust eventually melting within the mantle. Plate movements on convergent boundaries, causing earthquakes, may originate deep within the crust due to this vertical movement. Location *B* is on a transform plate boundary. Here, as shown by the arrows, the Pacific Plate and the North American Plate are sliding or slipping by each other. When movement occurs on a transform plate boundary shallow earthquakes may be generated since there is more horizontal movement.

81. Acceptable answers include, but not limited to:
A is located at a plate boundary, and *D* is not located at a plate boundary.
or Crustal plates are colliding at *A*, and no plate collision is occurring at *D*.
or Location *A* is on the Pacific Ring of Fire.

Explanation: Locate positions *A* and *D* on the Tectonic Plates chart in the RT. Location *A* is on a convergent boundary plate boundary. Here an ocean plate is subducting producing the Aleutian Trench. At plate boundaries, earthquakes can be expected frequently. Location *D* is located near the center of a plate. Since this area is located far from any plate boundary, few earthquakes would be expected.

82. Acceptable answers include, but not limited to:
E is located above a mantle hot spot.
or *E* is Canary Islands Hot Spot
or *F* is near the center of a tectonic plate.

Explanation: Open to the Tectonic Plates chart in the RT and locate positions *E* and *F*. Position *E* is very close to the Canary Islands Hot Spot. Hot spots are areas that have frequent volcanic eruptions producing volcanoes. Position *F*, is located in the middle of the Eurasian Plate and the Tectonic Plate chart shows no hot spot symbol for this location, thus it would be unlikely to have volcanoes.

83. Acceptable answers include, but not limited to:
    Location *G* is the ridge and is presently forming, while *H*
    was at the ridge in the past.
    or Location *H* is moving away from the new crust forming in the region at *G*.
    or The youngest ocean-floor bedrock is at the mid-ocean ridge.
    or The plates are diverging at the Southeast Indian Ridge.

    Explanation: Locate position *G* and *H* on the Tectonic Plates chart in the RT.
    Location *G* is on a divergent (spreading) plate boundary. Here, as the plates
    spread, magma rises, cools and produces the youngest igneous rocks while
    making new oceanic floor. As one moves further away from the ridges the age
    of the ocean floor increases, since these rocks (ocean floor) have been moving
    slowly away from the spreading ridge. This is a proof of sea-floor spreading.
    Thus, position *H*, being further from the ridge must be older than any
    position closer to the ridge where the younger rocks are.

June 2006

1. 3   The moon revolves around the Earth; it is our natural satellite. It is much closer than our Sun and all other planets.

2. 2   The altitude of Polaris (the north star) measured up from the horizon is equal to the observer's latitude. The diagram shows that Polaris is 66.5° from the horizon, this makes the observer latitude, 66.5°N.

3. 3   Open to the Relationship of Transported Particle Size to Water Velocity chart in the RT. The smallest boulders are represented on the 25.6 cm dashed line. At the intersection of this dashed line and the darken graph line, move directly down to the Stream Velocity axis. This intersects at the 300 cm/sec stream velocity position.

4. 4   Open to the Geologic History of NYS chart in the RT, and on the lower left side is the Archean Era, the oldest era of the Precambrian. In this area extending into the Proterozoic Era is written "Transition to atmosphere containing oxygen." From this statement it can be assumed that the Archean Era had very little oxygen.

5. 1   The moon's gravitational pull on the oceans is the major cause of tides. The Sun, being further away has a smaller effect on tides. At position 1 and 5 the combined gravity of the Moon and Sun causes the greatest difference between high and low tides. These are called spring tides and have nothing to do with the seasons.

6. 1   Open to the Luminosity and Temperature of Stars chart in the RT and locate the Red Dwarfs. Their temperatures are relatively cool (around 2500°C to 3500°C), while their luminosity is relatively low, under .01.

7. 2   Our solar system is located in an arm of the Milky Way Galaxy. All of the billions of galaxies are located within our Universe.

8. 3   The swinging pendulum appears to slowly change direction knocking over the pegs. The pendulum is not changing direction as it swings, rather the Earth is rotating under it causing the apparent change of direction. The Foucault pendulum is a proof of the Earth's rotation.

9. 2   On March 21, (first day of spring) the Sun's direct rays (vertical rays) strike the equator. Three months later, on June 21, the Sun direct rays have moved 23.5° north to the Tropic of Cancer (23.5°N). Thus, on June 21 the Sun would be 23.5° higher in the sky as seen for an observer at 42°N.

10. 2   A lunar eclipse occurs when the moon moves into the Earth's shadow. The correct order for a lunar eclipse is: Sun – Earth – Moon, which is shown in diagram B. Diagram D shows the correct order for a solar eclipse: Sun – Moon – Earth.

11. 4   The deflection or curvature of moving objects such as wind, and surface ocean currents is known as the Coriolis effect. This shift in direction is caused by the rotating Earth. Open to the Planetary Wind and Moisture Belts in the Troposphere and the Surface Ocean Currents charts in the RT and this deflection of planetary winds and ocean surface currents is shown by the curving arrows.

12. 3   Differences in air pressure will cause the atmosphere to move, generating wind. The larger the pressure gradient, the faster the wind will be. This occurs on Day 3 when a difference of 10.9 mb existed between the two cities.

13. 1   Open up to the Selected Properties of Earth's Atmosphere in the RT. Using the km scale, the troposphere starts at sea level and ends approximately 13 km in altitude. The cloud is located totally within the troposphere.

14. 4   As cold air enters the house it will sink being denser than warmer air. This colder air will displace the warmer air upward, being less dense, forcing some of the warmer air out from the top of the window.

15. 2   Open to the Surface Ocean Currents in the RT and locate the Gulf Stream Current in the Atlantic Ocean. The given key shows it is a warm current flowing northeastward toward Europe.

16. 3   Weathering is the breaking down of the lithosphere by chemical and physical means. Erosion is the transportation of weathered sediments to a new location by agents of erosion. Answer 3 is an example of erosion, while the others are examples of weathering.

17. 1   Carbon dioxide (CO_2) is recognized as a greenhouse gas. Greenhouse gases absorb infrared radiation causing an increase in atmospheric temperature. As the concentration of CO_2 increases, the amount of infrared radiation absorbed will increase, this is a direct relationship.

18. 3   Position X is being rotated into the night. This happens around 6 p.m.

19. 4   Open to the Planetary Wind and Moisture Belt in the Troposphere chart in the RT and locate the South Pole. The arrows show that air tends to sink over this polar region. This very cold sinking air can hold little moisture, thus this area is labeled "Dry". The same is true for the North Polar region.

20. 2 All conditions between X and Y are the same, except for the gradient (slope). An increase in gradient would result in an increase in stream velocity. On a topographic map close contour lines indicate a steep gradient as shown by diagram 2.

21. 3 Open to the Rock Cycle and the Sedimentary Rock charts in the RT. The Rock Cycle chart shows that compaction and cementation of sediments is needed to produce a sedimentary rock. The Sedimentary Rock chart shows that sandstone is an inorganic land-derived sedimentary rock having a clastic texture.

22. 4 Open to the Tectonic Plates chart in the RT. These plate boundaries are where major earthquakes are more likely to occur. The west coast of South America is where the Nazca and the South American Plate are converging.

23. 2 Index fossils are extremely helpful in correlating, or matching, specific sedimentary layers (such as the Devonian-age strata) throughout the world. The conditions for a fossil to be recognized as an index fossil are: they must be abundant, be widespread and lived a relatively short time compared to the geologic time scale. Specific index fossils are shown on the top of the Geologic History of NYS chart in the RT.

24. 2 Open to the Earthquake P-wave and S-wave Travel Time chart in the RT. Go to the 4000 km (4×10^3) epicenter distance position and move upward until it intersects the P–wave travel line. This intersection occurs at the 7-minute line. Continue upward and it intersects the S-wave travel line at 12 minutes and 40 seconds (12:40). The difference, or separation time, between these two lines is 5:40. If the P-wave arrived at 10:00:00, then the S-wave would arrive at 10:05:40.

25. 1 Open up to the Inferred Properties of Earth's Interior chart in the RT, and locate the density chart on the right side. Using the given density ranges, the mantle is less dense then both cores, but more dense than the crust.

26. 4 Open up to the Tectonic Plates chart in the RT and find the Divergent Plate Boundary key. This spreading boundary occurs over the Iceland Hot Spot.

27. 4 Open to the Geologic History of NYS chart in the RT. Using the Life on Earth column, the earliest fish appeared in the Cambrian Period. As one moves upward the new-life forms become younger in age. Insects appeared next in the Silurian Period, then amphibians in the Devonian Period, and finally reptiles appeared in the Carboniferous Period.

28. 3   Fossils are found almost exclusively in sedimentary rocks. Many sedimentary rocks are produced by compaction and cementation of sediments under-water (see the Rock Cycle chart in the RT). Open to the Geologic History of NYS chart in the RT and located at the top is this fossil - letter A. In the Time Distribution of Fossils column, letter A is positioned on the darkened trilobite bar at the Cambrian Period. Trilobites lived in shallow seas making them a marine animal.

29. 1   Open to the Geologic History of NYS chart in the RT and located on the top right is fossil W, a maclurite. In the Time Distribution of Fossil column, locate fossil W on the gastropod darkened bar. It existed during the Ordovician Period. Now, open to the Generalized Bedrock Geology of NYS chart and find the Ordovician Period key. Notice the surrounding bedrock by Watertown is Ordovician in age in which maclurites would be found. Using the Landscape Regions of NYS chart, this area of Ordovician-age rocks is within the Tug Hill Plateau landscape region.

30. 1   The contour interval, or spacing, for this topographic map is 20 m, making location $X = 380$ m and location $Y = 300$ m. This makes 80 m for the change in field from X to Y (see the Gradient equation found on the first page in the RT). Using the given scale, the distance from X to Y is 2 km. Substituting: $G = 80$ m/2 km or $G = 40$m/km.

31. 2   Open to the Relative Humidity chart in the RT. The wet bulb is 15°C and the dry bulb is 20°C, making a difference of 5°C. Go down the 5°C column until it intersects the Dry-Bulb temperature 20°C. At this intersection point is the answer of 58%.

32. 4   Open up to the Tectonic Plates chart in the RT and locate the key area at the bottom. The given diagram shows the slippage of two plates at a transformed plate boundary. The Tectonic chart shows this plate boundary exists in California, producing the famous San Andreas Fault.

33. 2   The Gulf of Mexico is a source area for mT air masses. These mT air masses are very moist (m = maritime = wet) and warm (T = tropical = warm). Central Canada is a source area for cP air masses. cP air masses are dry (c = continental = dry) and cold (P = polar = cold). All other maps show wrong origins of air mass. (All air masses symbols are found in the Weather section of the RT.)

34. 4   Differences in abrasion is attributed to the hardness of the mineral. The softer the mineral is, the faster it would abrade. To compare the hardness of these minerals use the Properties of Common Minerals chart found in the RT. Galena having the lowest hardness (2.5) of the given choices would experience the most abrasion.

35. 1   The underlying bedrock structure of a plateau will show horizontal sedimentary strata. These elevated layers become eroded, or dissected, over time and may produce large hills with deep valleys as shown by diagram 1.

36. 2   The intersection point of 500°C and 2000 atmospheres of pressure falls in the shaded area of the mineral Andalusite. All other given choices are outside this shaded area.

37. 4   On the top of the diagram is given the chemical formula for these three minerals, Al_2SiO_5 making them members of the silicon family. Open to the Properties of Common Minerals and find potassium feldspar. It has a similar formula, $KAlSi_3O_8$, making it also a member of the silicon family.

38. 3   In the metamorphic process the addition of heat and pressure causes a change. Open to the Rock Cycle chart in the RT and here it shows these two conditions for metamorphism.

39. 4   The periods (time) of revolution, the orbit around a central object, and rotation, the spin around the axis, are known to astronomers. These periods repeat in a cyclic manner making them very predictable.

40. 3   Open to the Solar System Data chart and locate the Earth's moon row. Here it shows that the period of rotation and revolution for the Moon are very close, just over 27 days. This is why we always see the same side of the Moon.

41. 4   Letter *D* symbolizes the orbit, or path, of the Moon around the Earth. As the Moon revolves around the Earth, we see more or less reflected light from the moon. This changing amount of light causes the different phases of the Moon.

42. 2   The original radioactive sample had 24 squares. After one half-life, half of the radioactive substance will decay into a new substance. In this case, after the first half-life 12 squares would still be radioactive, while 12 squares would have decayed, this is shown in the diagram. The second half-life would cut the 12 remaining radioactive squares by half again, leaving 6 white radioactive squares and shading in 6 more decayed squares, making a total of 18 shaded (decayed material) squares.

43. 1   Open to the Radioactive Decay Data chart on the front cover of the RT. Carbon-14 has a half-life of 5,700 years. In the instructions, the half-life of this sample is given as 5000 years.

44. 1   A typical watershed is a large drainage area, consists of hundreds of small streams/creeks called tributaries that carry water to a main river. This picture of Esopus Creek is indicating that it is a tributary of the Hudson River, part of the Hudson River watershed.

45. 2   The steeper the gradient (slope) the faster the water will flow causing much down cutting eventually producing a V-shaped straight channel - Diagram *B*. As the gradient becomes less, the water starts to erode the channel more laterally, side to side - Diagram *C*. On relatively flat surfaces the stream channel will show much meandering producing a broad flood plain – Diagram *A*.

46. 4   Water will erode sediments from upstream bedrock carrying these sediments a long distance. As sediments travel downstream they become rounder and smaller by abrasion. On relatively flat areas such as a floodplain shown in Diagram *A*, the steam's velocity decreases causing much deposition of these upstream weathered sediments.

47. 2   Due to the tilt and revolution of the Earth around the Sun, insolation (incoming solar radiation or simply sunlight) changes with the seasons and latitude. The graph clearly shows a major difference in insolation from the Polar region (90° N) and the equatorial region throughout the year. Remember, latitude degrees are followed with N and S, while longitude degrees are followed with E and W.

48. 3   On March 21 and September 22, the equinoxes, the Sun's direct, or vertical rays producing the most intense insolation (sunlight), fall on the equator. Due to the tilt and the revolution around the Sun, the Sun's direct rays will move north to the Tropic of Cancer, 23.5° N, on June 21. Now this area will have more intense insolation then the equator area.

49. 3   Due to the tilt and revolution of the Earth around the Sun, the Sun's direct rays move between the latitude lines of 23.5° N to 23.5° S. This will also cause the height of the noon Sun to change throughout the year. From a Polar view, the Sun remains relatively low in the sky and does not set from March 21 to September 22. After September 22, the Sun dips below the horizon for the next 6 months. Without the Sun's rays, the insolation value would be zero.

50. 1   Tube *B*, having larger particles, causes water to flow faster with less water being retained, or held back, compared to Tube *A*. These two conditions are correctly shown in Data Table answer 1. Since both tubes are filled to the same height and have beads of different but uniform sizes (sorted sizes), the water amount required to fill pore spaces would be the same. Unsorted sediments (having a mixture of sizes) would have less pore spaces.

Part B–2

51. Answer: Any value above 20 ft but below 30 ft.

Explanation: Contour lines show elevation on a topography map. As shown by the diagrams Point *A* is situated on a hill between the two 20-foot contour lines and the 30-foot contour line to the west. Thus, Point *A* must have an elevation between these values.

52.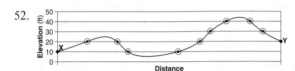

Explanation: Place the edge of a piece of paper along the line *X-Y* on the topographic map. Make a mark on the edge of the paper where each contour line intersects the paper. Place the edge of the paper, with the marks, along the bottom line of the profile graph. Move vertically up the graph above each mark to its correct contour value and put a dot at the proper elevation. Connect the points making sure you extend your line below the lowest points to show the depression and above the highest points to show the two hills.

53. Acceptable responses include, but are not limited to:
A is slower cooling than *B*.
or *B* is faster cooling than *A*.
or Intrusive rock forms from molten rock that cools slowly.
or Extrusive rock forms from molten rock that cools rapidly.

Explanation: Magma, inside the earth, cools much slower than lava, on the surface of the earth. The slower molten rock cools, the larger the crystals will be. This is why intrusive igneous rocks have coarser texture than extrusive igneous rocks (see the Texture area in the Igneous Rock chart in the RT).

54. Answer: 1 mm to 10 mm

Explanation: Open to the Igneous Rock chart in the RT and locate the Texture area. Go down to the coarse texture row, where the grain size is given as 1 mm to 10 mm. Remember, this texture size is referring to the size of the crystals that grew from the magma.

55. Acceptable responses include, but are not limited to:
obsidian   *or* basaltic glass
or pumice   *or* vesicular basalt glass

Explanation: Rock *D* must be an extrusive noncrystalline igneous rock as shown by the given chart. Open to the Igneous Rock chart in the RT and locate the Grain Size column and then locate the Noncrystalline rows. Moving to the left shows the four rocks that are placed in this classification.

56.

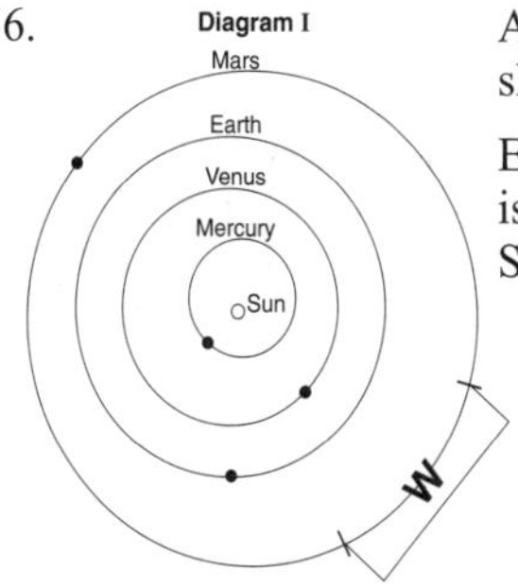

Diagram I

(Not drawn to scale)

Answer: Letter *W* must be placed within the area shown indicated by the bracket.

Explanation: The Sun's gravitational force on Mars is the weakest when Mars' orbit is farthest from the Sun. Area *W* indicates this section.

57. Answer: Circle Saturn and Jupiter.

Explanation: The Jovian planets are the giant gaseous planets in our Solar System. They are: Jupiter, Saturn, Uranus, and Neptune. Of these, the two largest are Jupiter and Saturn. This is shown in the Solar System chart in the RT in the Equatorial Diameter column.

58. Answer: 84 years or 84.0 years. Correct units must be included in the answer.

Explanation: One complete orbit of a planet would be its period of revolution. Open to the Solar System chart in the RT. Locate the Period of Revolution column and go down to Uranus.

59. Acceptable responses include, but are not limited to:
The orbits are elliptical or oval shaped.
or The orbits are nearly circular.

Explanation: All planets revolve around the Sun in elliptical orbits, which are oval in shape. They do not revolve around the Sun in circular orbits. Each planet's orbit has an eccentricity number shown in the Solar System Data chart in the RT. The larger the eccentricity number, the more elliptical the planet's orbit will be.

60. Acceptable responses include, but are not limited to:
Sometimes Pluto is closer to the Sun than Neptune is.
or Part of Pluto's orbit is sometimes located within Neptune's orbit.

Explanation: Orbital speed is related to the distance from the Sun. Due to the gravitational pull the Sun exerts on the planets, the closer a planet is to the Sun, the greater the orbital speed will be. Pluto is usually the farthest planet from the Sun, thus having the longest period of revolution (time to make one complete orbit). Due to Pluto's large eccentricity (see explanation 59), Pluto's orbit is highly elliptical in shape compared to the other planets. This causes Pluto to be closer to the Sun, at times, than Neptune. When this occurs, Pluto's orbital speed will be faster than Neptune's.

61. Answer: Pleistocene Epoch.

Explanation: In the instructions it states that the given map shows the advances of the last continental ice sheet. Open to the Geologic History of NYS chart in the RT and locate the Important Geologic Events in NY column. The top section states, "Advance and retreat of last continental ice". Move directly to the left in this row to the Epoch section. It lines up within the Pleistocene Epoch.

62. Acceptable responses include, but are not limited to:
Glacial sediment is unsorted.
or Piles of mixed sediment sizes.

Explanation: A glacier acts like a large bulldozer, scraping, dragging and carrying all sizes of sediment as it advances. When it melts and retreats, these sediments are released (deposited) in an unsorted manner, all mixed together.

63. Answer: southwest (SW) *or* south southwest (SSW)

Explanation: Open to the Generalized Landscape Regions of NYS chart in the RT and locate the Catskills. Knowing this position, the given Ice sheet map shows that the Hudson-Champlain Lobe moved over the Catskills in a southwest direction. Remember, the direction of the glacier flow is where it is going to, not where it came from.

64. Acceptable responses include, but are not limited to:
parallel scratches
or grooves
or striations
or orientation of glacial features, such as drumlins and lateral moraines

Explanation: A glacier acts like a giant bulldozer carrying large amounts of unsorted sediments. Some of these sediments are large boulders that are suspended in the ice. As the ice slowly moves some of these boulders come into contact with the solid bedrock, scratching and grooving the bedrock. These marks will be made in the direction that the ice is moving. Also, when these giant ice sheets melt, sediments are released producing small hills called drumlins or sediment deposits called lateral moraines that are oriented in the direction the ice was flowing.

Part C

65. Acceptable responses include, but are not limited to:
Water droplets form on the surfaces provided by the salt and dust particles.
or Salt and dust particles are condensation nuclei, allowing the water vapor to change into liquid drops, forming clouds.

Explanation: Scientist discovered that all cloud droplets (floating dew drops) have some sort of dust or salt or pollution particles attached to them. These are called condensation nuclei and must be present to form cloud droplets. It is around these particles that water vapor can condense when the dew-point temperature has been reached.

66. Acceptable responses include, but are not limited to:
Dust particles can be blown into the atmosphere by winds.
or a volcanic eruption   *or* a forest fire   *or* space impacts

Explanation: All of the above examples emit tons of ash and particles into the atmosphere yearly. These dust particles may become condensation nuclei for the formation of clouds. Past evidence indicates major impact collisions of meteorites and asteroids with the Earth have occurred.These space impacts can throw up enormous amounts of particles into the atmosphere. In severe cases the ejected ash can reduced insolation (sunlight) causing climate changes.

67. Acceptable responses include, but are not limited to:
The air on the western slopes of the mountains is rising.
or The valleys are located on the eastern side of mountain ranges
    where air is sinking.
or Air is warmed by compression as it descends the mountain slopes,
    so relative humidity decreases.

Explanation: The valleys, having less than 20 inches of yearly precipitation, are on the leeward side of the mountains. On the leeward side of the mountains, the relatively drier air (much moisture was released on the windward side) gets compressed as it flows downward into the valley. This causes the air to increase in temperature resulting in a decrease in the relative humidity. This results in few clouds and little precipitation producing a desert or arid climate on the leeward side.

68. Answer: 134 in ± 4 in.

Explanation: Find the four points (dots) on the mountains within the Coastal Mountain Ranges. For each point, read directly down to the bar graph shown, then read over to the Yearly Precipitation scale. If done correctly, the first point receives a yearly precipitation amount of 33 inches, the second 52 inches, the third 17 inches and the forth 33 inches. This adds up to be a total yearly precipitation of 134 inches.

69. Acceptable responses include, but are not limited to:
The Sierra Nevada Mountain Range is higher in elevation.
or Higher elevations have lower temperatures.
or Expansional cooling increases with higher mountains.

Explanation: As the altitude increases within the troposphere, the temperature decreases. This inverse relationship is shown on the Selected Properties of Earth's Atmosphere chart in the reference table. This occurs, since rising air expands due to the decrease in atmospheric pressure. This expanding air produces a cooling effect on the atmosphere.

70.

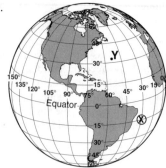

Answer: See position of **X**.

Explanation: Latitude lines run E to W, but are measured N and S from the equator. All southern latitude readings are below the equator. 30° W is a longitude reading. Longitude lines run N and S but are measured E and W from the Prime Meridian, which is the O° line that runs through England. Use this information to accurately place your **X** as shown in the above answer.

71. Answer: 15°/hr.

Explanation: The Earth rotates 15° per hour no matter where you are. Based on this rotational speed of 15°/hr, time zones were set-up for every 15° of longitude.

72.

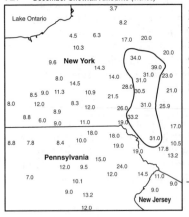

Map 1
December Snowfall Amounts (inches)

Answer: See map.

Explanation: The 30.0 inch snowfall isoline would connect all areas that received 30 inches of snow. Carefully draw your line estimating where 30.0 would be positioned by studying the given numbers. Remember, isolines can never cross.

73. Acceptable responses include, but are not limited to:
Residents should have bought extra supplies such as food, milk, and water.
or Residents should have gotten their battery-powered radio, flashlights, and candles ready.
or Residents should have made sure emergency generators work.
or Residents should have checked that they had enough fuel, oil, or wood for heat to last several days.

Explanation: In forecasted weather emergencies, fresh water, a first-aid kit, and a portable battery operated radio are essential items. All of the above given answers are helpful steps in preparing for a snow emergency. Also, in an upcoming major snowstorm, all Earth Science students should bring home their textbook to study during your snow day.

74.

Weather Conditions	Description
present weather	snow
wind direction from	northeast
wind speed (knots)	10 ± 2
relative humidity (%)	100

Answer: See chart.

Explanation: Using the General Bedrock Geology chart in the RT, locate Buffalo in the western section of NYS. On Map 2, locate Buffalo's weather station model. Open to the Weather Map Symbols in the RT. Using this chart, correctly identify the given weather conditions at Buffalo. The wind direction is always from where the wind is coming from. The relative humidity is not placed on the station model. But, when the dew point temperature and the air temperature are the same (which is true for Buffalo) the relative humidity will always be 100 %.

75. Acceptable responses include, but are not limited to:
Winds are counterclockwise and inward toward the low-pressure center.
or Winds are counterclockwise.
or Winds are blowing toward the center of the low.

Explanation: Around a Low pressure system, surface air moves in a counter-clockwise motion and inward. On Map 2, this motion of air around the **L** is shown by the wind direction shafts of the given station models.

76. Acceptable responses include, but are not limited to:
east *or* northeast *or* north northeast *or* east northeast

Explanation: U.S mainland weather generally moves towards the NE. This occurs due to our planetary wind belt called the westerlies. Here strong winds, including the jet streams, generally blow from the southwest to the northeast, moving air masses, (pressure systems) fronts, and our general weather in this direction. (Note, there are exceptions to this rule.)

77. Answer: quartzite *or* hornfels.

Explanation: Point *A* is positioned at the contact metamorphism area within the sandstone region. Contact metamorphism produces metamorphic rocks due to heat of molten rock (lava or magma), in this case, the heat from the basaltic intrusion (magma). Open to the Metamorphic Rock Chart and locate quartzite. In the comment section, it states "metamorphism of quartz sandstone". The chart also shows that contact metamorphism produces hornfels.

78. Acceptable responses include, but are not limited to:
The fault displaced the intrusion.
or The fault has cut across the preexisting basalt intrusion.

Explanation: Fault lines are always younger than the geologic structures (layers, intrusions, ect.) they cut. In this case, the magma intruded the layers and cooled, producing an igneous basaltic intrusion. Sometime after the breccia (see rock key) was deposited, an earthquake occurred displacing (cutting) the intrusion along the fault line.

79. Answer: Formed first: limestone
                Formed second: breccia
                Formed last: basalt

Explanation: Usually the oldest layers are on the bottom and the youngest on top. This makes limestone older than breccia. Yet, the basaltic intrusion is the youngest event of the three. The basalt intruded or entered both the limestone and the breccia. Both must have already been deposited there for the intrusion to cut through it. Also, contact metamorphism is shown on both layers; this can only happen when the layers are present. Contact metamorphism occurs when rocks are changed by the heat of molten rocks (lava/magma).

80. Answer: 0.006 cm *or* .006 cm.

Explanation: Open to the Sedimentary Rock Chart in the RT and locate the grain size for siltstone. Here it shows the size range of particles that are considered to be silt. The smallest accepted size is 0.00004 cm and the largest particle size for siltstone is 0.006 cm.

81. Acceptable responses include, but are not limited to:
quartz *or* plagioclase feldspar *or* biotite
or amphibole (hornblende) *or* pyroxene

Explanation: Open to the Igneous Rock Chart in the RT. Locate andesite in the middle of the chart. Move directly down to the mineral chart. Staying within the boundary of the dashed lines, we come to the minerals that are found in this rock. Notice, that one can expect to find a very small percentage of quartz, but a large amount of plagioclase feldspar.

82.

(Not drawn to scale)

Answer: Placing the ▲ on Earth's surface above the magma chamber.

Explanation: The reading implies that volcanoes usually occur above magma chambers. Magma chambers contain the necessary molten rock for volcanism. As pressure increases within the chamber, the molten rock is forced upward to the surface of the earth producing a volcano or a lava flow.

83. Answer: See above diagram. All three arrows must show the correct directions.

Explanation: When an oceanic plate collides with a continental plate, the denser oceanic plate will subduct, sliding downward under the continental plate to the mantle. When two plates collide a convergent boundary is produced. Correct positioning of the arrows, indicating its motion, will show the plates coming at each other, producing this subduction zone.

June 2007

1. 4 A swing pendulum will appear to change direction over a short period of time (within hours). This apparent shift in direction of the pendulum is caused by the Earth's rotation. The Foucault pendulum is a proof of the Earth's rotation.

2. 3 According to the Big Bang theory, distant galaxies are still moving away from us and each other at tremendous speeds. Evidence of this expanding universe is obtained by studying the redshift of spectral lines of distant galaxies. These spectral lines will be moved or shifted toward the red end of the spectrum if objects, such as distant galaxies, are moving away from the observer.

3. 2 On the windward side of a mountain (letter *A* side), moist air rises, expands, and cools to the dewpoint temperature, producing clouds and eventually rain. This makes the climate on the windward side cool and humid. On the leeward side (letter *B*), the drier sinking air becomes compressed, which increases the air's temperature. This produces an arid or desert climate.

4. 1 The equator receives direct sunlight or almost direct sunlight all year round. Direct sunlight (insolation) are rays that strike the Earth at 90° and cause a rapid increase in temperature, which the equator is famous for. Due to the Earth's tilt of the axis, the North and South Polar regions receive only indirect sunlight, which has little heating ability. For half the year these areas receive no sunlight, being tilted away from the Sun.

5. 3 Open to the Surface Ocean Currents map in the RT. At 20° S, the western coast of South America is being influenced by the cool Peru Current (see key), while the eastern coast is being influenced by the warm Brazil Current. Ocean currents have a major effect on the climate of coastal areas.

6. 2 Open to the Electromagnetic Spectrum in the RT. The right side of this chart is where the longest wavelengths are found, while the left side shows the shortest wavelengths. Radio waves are the longest electromagnetic waves, having a wavelength that may be longer than 10^3 cm. Remember, the electromagnetic spectrum is composed of different wavelengths, but all of them travel at the same speed, the speed of light.

7. 2 Evaporation (L→G) is greatly favored by dry air, which will have a low relative humidity value (RH). No additional evaporation will occur when the air is saturated, having an RH of 100%. A breeze quickens evaporation by moving evaporating molecules away from the water source. At higher temperatures, the water molecules will have more energy, causing faster molecular motion, which produces faster evaporation.

8. 4   Open to the Selected Properties of Earth's Atmosphere in the RT. The Water Vapor chart shows that all the water vapor is contained in the first layer of the atmosphere, the troposphere. This is the reason that almost all of the Earth's weather occurs in the troposphere.

9. 3   Capillary action causes water to move upward against gravity within small porous substances. This action has caused the water to move upward in the dry clay pot to level B. Capillary action is favored by small sediment size, which has small pore spaces.

10. 3   If there is a strong air-pressure gradient, isobars on a weather map will be close together. This rapidly changing pressure field will result in strong winds, as air moves from the high pressure to the low-pressure area.

11. 2   When the dewpoint temperature and air temperature become closer together, the relative humidity (RH) increases. When these two temperatures are the same, the RH value is 100% and condensation will begin (G $\rightarrow$ L). In the atmosphere this produces clouds, increasing the chance of precipitation.

12. 1   Open to the Average Chemical Composition of Earth's Crust, Hydrosphere, and Troposphere chart in the RT. The Crust column shows that by mass, oxygen (O) is 46.40% and silicon (Si) is 28.15% of the total composition of the crust. These two elements readily join to produce the mineral quartz, SiO_2.

13. 3   The instrument shown is a hydrometer, containing a wet-bulb and a dry-bulb thermometer. Open to the Relative Humidity chart in the RT. The difference between the dry-bulb and wet-bulb is 8°C. Locate this difference at the top of this chart. Staying in this column, moving downward, stop at the Dry-Bulb temperature of 18°C. At this intersection position, the given RH is 33%.

14. 4   Open to the Properties of Common Minerals chart in the RT. In the Distinguishing Characteristics column, move down to Halite. The given diagram shows the 90° cubic cleavage planes of this mineral. Continue down in this column to Pyroxene. Here it states "cleaves in 2 directions at 90°."

15. 2   When water slows down, deposition of sediments occurs. The largest sediments get deposited first, while the smallest sediments are deposited last. This produces much sorting of sediments (by size) and is a major clue of water-transported sediments. Glacier deposition is highly unsorted, in which large and small sediments are mixed together. As the ice melts, the locked up sediments of all sizes fall out, causing this unsorted manner.

16. 2   Open to the Generalized Bedrock Geology of NYS. Go to the Key area and locate the Devonian and Silurian key. Just to the right of these two keys are given the rock types that one can expect to find in these age bedrocks. It also states: "*Silurian also contains salt, gypsum and hematite.*" Syracuse is located on Silurian bedrock and salt is found there.

17. 4   A feature of wind erosion, especially in arid climate areas, is the development of sand dunes. As sand is transported by wind, it builds up a dune that develops a characteristic shape having a steep slope on one side and a gentle slope on the other side (see picture). From this picture the prevailing winds, blowing from the right, move the sand grains up the gentle slope until they fall (being deposited) down the steeper slope of the dune. In time, this causes the sand dune to slowly migrate or move.

18. 2   Lava will experience quick cooling since it is on the Earth's surface, while magma, being insulated by surrounding rocks, will cool much slower. Fine-grain and glassy textures are formed by the relatively fast cooling lava, while magma produces a coarse texture. These environments and texture types are given in the Igneous Rock Identification chart in the RT.

19. 1   At a convergent plate boundary, a subduction zone is produced when a continental plate overrides an oceanic plate (see Key area in Tectonic Plates chart). The denser oceanic plate will sink or subduct into the mantle producing a trench within this subduction zone. Open to the Inferred Properties of Earth's Interior chart and the densities of the continental and oceanic plates are given at the top right of this chart.

20. 3   Open to the Geologic History of NYS chart and locate the Life on Earth column. Here it documents the great diversity of life that the Earth has supported, of which many species are now extinct. The chart mentions extinction at the end of the Paleozoic and the Mesozoic era. Mass extinction has been documented by fossil evidence many times throughout geological history.

21. 1   Open to the Geologic History of NYS chart in the RT. At the top of this chart are diagrammed important index fossils found in NYS. Letter Q shows the fossil tree – Aneurophyton. Fossilized trees indicate that a forest environment once existed in that area. All other choices are marine organisms and indicates that a sea once covered the area.

22. 3   Mass movements, which include rockslides, landslides, slump etc., are when large section of the Earth's crust moves or separates under the influence of gravity. These movements are accelerated by heavy rainfall and strong vibrations. The diagram shows a large section of sandstone that has broken off as the weaker shale experienced a mass movement.

23. 4   Open to the Generalized Landscape Regions of NYS. Locate the Newark Lowlands in the southern part of NYS. Now open to the Generalized Bedrock Geology of NYS and locate where the Newark Lowlands would be on this chart. If done correctly, the given Geological Period Key shows this region consists of bedrock from the Late Triassic and Early Jurassic Periods. To the right of this key is given the bedrock composition of red sandstone and conglomerates.

24. 2   Shale is a sedimentary rock that was made by the burial, compaction and cementation of clay. This information is given in the Sedimentary Rock chart and the Rock Cycle chart in the RT. Remember to use the Rock Cycle chart for the processes that produce the different rock types.

25. 1   The Universe is all of space. Within the universe are billions of galaxies that contain billions of stars. Solar systems consist of a center star (like our Sun), in which smaller planets revolve around. Diagram 1 shows this large to smaller sequence of symbols.

26. 4   Dark surfaces absorb insolation (sunlight) much better than light colors. A rougher surface will trap more insolation than a smooth surface of the same size. A smooth, light color surface will reflect the most insolation.

27. 1   The Sun rises in the east. But, due to the tilt of our axis and the motion of revolution and rotation, the Sun rises at slightly different positions in the east throughout the year. During the summer months, the Sun rises north of east, and during winter months the Sun rises south of east. On the equinoxes, the Sun rises due (directly) east, being over the equator. Answer 1 shows the correct dates of sunrise paired with their given position.

28. 3   The seasons in the Southern Hemisphere are opposite that of the Northern Hemisphere. This is caused by the tilt of the Earth's axis and the motion of revolution. When we are having summer, the Southern Hemisphere is experiencing winter. Graph 3 shows the coldest months are during June, July and August, and the warmest months are during our winter months. So this graph must represent the average monthly temperature for the Southern Hemisphere.

29. 2   Open to the Weather Map Symbols in the RT. As shown in the given station model, the wind direction is a shaft that is drawn from the direction of the wind into the station model circle. The wind speed is shown by feathers, where a 25 knot wind would have 2½ feathers on the wind direction shaft. Knowing this information, answers 1 and 2 have the wind direction and speed correct. Barometric pressure readings are abbreviated on a station model. This is done by dropping the 9 or 10 from the front of the reading and dropping the decimal. Using this rule, a reading of 1023.7 mb becomes just 237, correctly shown in answer 2.

30. 4   Solving for density of each sample using the density equation $d = m/v$, all samples have a density of 2.0 g/cm^3. The same density value would be expected since the samples are composed of the same mineral. On a graph, this relationship would be a straight line showing the density remains the same.

31. 2   Contact metamorphism occurs when magma or lava comes in contact with preexisting rocks and the heat of the molten rock causes a metamorphic change to these rocks. The diagram shows contact metamorphism of sandstone, which will change into the metamorphic rock quartzite. Locate Quartzite in the Metamorphic Rock chart and to the right of it, in the Comments section, it states "metamorphism of quartz sandstone." Information on contact metamorphism is given in the Metamorphic Rock chart in the Hornfels comment section.

32. 1   The youngest rocks layers are normally found at the top, but intrusion may be younger than the top layer. Intrusions will cause contact metamorphism on the rock layer above it, while an extrusion will not produce any contact metamorphism on the rock layer above it (see explanation no. 31). The diagrammed igneous rock did not cause any contact metamorphism of the shale, thus it must have been an extrusion in which the lava surfaced and cooled producing this igneous rock. Years later, clay sediments were deposited over the igneous rock and eventually underwent compaction and compression producing the sedimentary rock – shale.

33. 4   Open to the Tectonic Plates map in the RT. California is located on a transformed plate boundary, the San Andres Fault line (see the Key at the bottom). A Hot Spot is located at Yellowstone Natural Park. Hot Spots are volcanically active areas capable of producing major earthquakes. These areas are singled out as potentially producing a future major earthquake in the western part of United States.

34. 3   Open to the Generalized Bedrock Geology of NYS map in the RT. On this map locate the city of Plattsburgh in the northeast section of the state. The position of this city is in the dark band that represents major to extreme earthquake damage could be expected. There is an active fault that runs through this area.

35. 3   Open to the Geologic History of NYS chart in the RT. Locate the time bar on the far left side of the chart. The gray area, representing the Precambrian contains close to 90% of the Earth's Geologic History. The Cenozoic era, the one we live in, has existed the shortest time of all eras.

36. 2   Polaris (the north star) is located directly over the northern axis of the Earth. At this location the altitude of this star is 90°. At the equator an observer would see Polaris at an altitude of 0°. Below the equator this star would not be able to be seen since the Earth's horizon bocks it. Remember, in the Northern Hemisphere, an observer's latitude is equal to the altitude of Polaris.

37. 3   Year round, the equator has just about has the same amount of hours of daylight, 12 hours. On June 21 at solar noon, location *A* would have direct sunlight. This marks the first day of summer for the North Hemisphere. On this date and since location *A* is north of the equator, this location would have more than 12 hours of sunlight. At times the North Pole is having 24 hours of sunlight. Remember that on the equinoxes, the Sun is directly over the equator and all places on the Earth experience equal length of day and night.

38. 2   On December 21, the Sun has reached its farthest southern position, being 23.5° S (location *B*). This is the first day of winter for the Northern Hemisphere and the first day of summer for the Southern Hemisphere. After this date for the next 6 months, the Sun's direct rays travel slowly northward to the 23.5° N latitude line.

39. 3   Stage 4 is the cooling that allowed protons and neutrons to form. To the right of this is given the time from the beginning of the universe, 10^{-6} seconds.

40. 1   The Big Bang Theory assumes that billions of years ago all the matter of the universe was concentrated into an extremely small position or point. In time, the unbalanced forces within caused a massive explosion, moving all matter outward at tremendous speeds. This expansion of the universe caused cooling and condensation of matter, producing the stars and planets. Scientists tell us this rapid expansion of the univers is still occurring.

41. 1   At Stage 3 the universe was at 10^{27}°C. This is a very, very high temperature. Following the temperatures to Stage 8, which is close to the present, continually decreasing to –270°C.

42. 4   Stage 7 first mentions stars and only Stage 8 mentions planets. Thus, our solar system, consisting of our Sun and planets, must have formed between these two stages.

43. 2   The heat energy inside our Earth has created convection currents that eventually rise upward to the lithosphere. When these convection currents reach the plates, they turn and move horizontally, affecting the plates by slowly moving them. These convection currents are suspected of being the driving force of plate tectonics. Open to the Inferred Properties of Earth's Interior chart located in the RT. On the cross-section diagram, the flow of these convection currents are illustrated.

44. 1   A warm, tropical climate will be found in the equatorial zone, being at or near the 0° latitude line. The map shows that during the Cambrian Period, Australia drifted to this position.

45. 1   The curving of the Earth's winds are being caused by the rotation of the Earth. Objects that travel long distances on the Earth's surface will experience this deflection or curvature, due to the rotation of the Earth. This is known as the Coriolis effect. Surface ocean currents also undergo this curvature as shown in the Surface Ocean Currents map.

46. 2 Open to the Generalized Bedrock Geology of NYS chart and it shows the latitude and longitude range of our state. Our latitude places us in the prevailing south-westerlies zone. These winds that primarily blow from the southwest to the northeast, move air masses that change our weather.

47. 3 The map shows that the trade winds converge (come together) at the equator. These winds are warm due to the direct sunlight the equatorial region experiences. At the equator (labled "Wet"), the warm air rises, cools and reaches the dewpoint temperature. At the dewpoint temperature, condensation occurs, building large expanding clouds, which produce much rainfall. This situation has produced the tropical rainforests found along the equatorial region.

48. 1 Streams will experience erosion on the outside of curves, while deposition will be dominant on the inside of curves of a stream channel. Water velocity increases on the outside of a curve causing erosion, while water slows-up on the inside curve channel, producing the deposition of sediments.

49. 3 When a stream enters a quiet body of water (lake, sea, etc.), deposition will occur, with the largest particles being deposited first. The smallest particles, as clay, will continue to flow farther out, eventually depositing far from the mouth of a stream.

50. 4 As the velocity of the water increases, more sediments will be transported by the stream. Water velocity can be increased by an increase in the stream's gradient, adding more water to the channel, or restricting the width of the channel. That is why streams move faster after a major rain storm or snow melt off.

Part B-2

51.

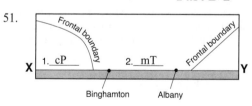

Explanation: Open to the Weather Map Symbols in the RT and locate the Air Masses section. These air mass symbols are based on moisture content (c = dry or m = wet) and the temperatures (T = warm or P and A = cold). Position 1 is behind the cold front, thus experiencing the cold air = P. Most of the cold air that enters NYS has its origin in central Canada, which is dry air = c. This makes this air mass a cP. Position 2 is behind the warm front, thus experiencing warm air = T. This air mass more than likely had its origin in the Gulf of Mexico making it wet = m. This makes this air mass a mT.

52. Answer: Binghamton

Explanation: This forecast is typical for a cold front during the summer months. Binghamton is the next city shown on the map that the cold front will pass through as it moves eastward. As the cold air collides into the warmer air, the warmer air (being less dense) will be forced to rise. This rising, cooling air will reach the dewpoint, causing condensation and eventually produces a line of thunderstorms.

53. Acceptable responses include, but are not limited to:
Move indoors.
or Do not use electrical equipment or telephones
or Do not stand under tall objects

Explanation: The above actions are common practices to follow when considering the safety aspects of an approaching thunderstorm. It is also a good practice to unplug electrical devices in a house, since a lightning strike often follows electrical lines, which can cause major damage due to a large voltage surge.

54. Answer: Cambrian Period

Explanation: Open to the Geologic History of NYS chart and locate the trilobite fossil shown in layer *C* with those diagrammed at the top. Layer *C* contains the trilobite – Elliptocephala (fossil *A*). This fossil is shown to have lived in the Cambrian Period by the positioning of letter *A* on the Trilobite bar in the Time Distribution of Fossils section within this chart.

55. Acceptable responses include any of the two answers below.
uplift *or* erosion *or* weathering *or* subsidence *or* deposition *or* burial

Explanation: An unconformity is diagrammed by a wavy line that represents a missing layer(s) due to erosion. This produces a time-gap in the geologic rock sequence. The normal sequence of events in producing an unconformity is weathering and erosion of rock layers, submergence (sinking of layers under water), deposition and burial of new sediments eventually creating new layer(s) over the eroded one. Crustal uplift brings the unconformity out of the water exposing it.

56. Acceptable responses include, but are not limited to:
widespread geographic distribution
or   short existence in geologic time
or   be abundant

Explanation: An index fossil is a fossil that is very helpful in the identification of a specific layer representing a specific geologic time period. For a fossil (the species) to be recognized as an index fossil, it must have lived a relatively short period of time, thus found in one specific geologic period strata (like Cambrian Period), be distributed widespread (over a very large area) and be abundant (making it easy to find).

57. Acceptable responses include, but are not limited to:
Carbon-14's half-life is too short.
or Not enough carbon-14 is left to measure.
or The fossils are too old.

Explanation: Open to the Radioactive Decay Data chart in the RT. The given half-life of C-14 is 5.7×10^3 or 5,700 years. This is a relatively short half-life compared to the other given radioisotopes. Due to its short half-life, objects older that 100,000 years (at the most) cannot be dated by C-14, since too little of this radioactive element remains.

58. Answer: pebble

Explanation: Using the metric ruler located in the referece tables found in this book, the largest sediment is 2.0 cm and the smallest is .5 cm. Open to the Relationship of Transported Particle Size to Water Velocity chart and on the right side of the graph it shows that pebbles have a size range of 0.2 cm to 6.4 cm.

59. Acceptable responses include, but are not limited to:
abrasion *or* weathering *or* erosion
or Particles were worn down as they were scraped along the bedrock.

Explanation: As the stream transports sediments, they will be worn down by the action of abrasion. As sediments bounce along the bottom of the stream channel, the edges become chipped and worn away, which results in a rounder and smaller shape.

60. Answer: shale

Explanation: A soft rock layer will undergo more erosion causing it to be cut back farther into the bedrock. The shale layer is cut down deeper than the other layers, producing the V-shape valley that is associated with stream erosion. The more resistant (stronger) layers would be higher up because they have not been eroded as much.

61. Acceptable responses include, but are not limited to:
A glacier forms a U-shaped valley
or Glaciers form U-shaped valleys and streams form V-shaped valleys.

Explanation: A glacier acts like a large bulldozer, slowly scraping and scouring out valleys as it moves through them. Their tremendous size and overall weight causes the valleys to become U-shaped. A valley in which water travels through tends to have a V shape.

62. Acceptable responses include, but are not limited to:
It shows banding. *or* The rock is foliated.
or The minerals are segregated into layers. *or* distortion

Explanation: Banding is a metamorphic process in which minerals are aligned (foliated) or segregated into a wavy pattern by the processes of heat and/or pressure associated with metamorphism. This information is given in the Metamorphic Rock chart, Comments section for the rock Gneiss.

63. Acceptable responses include, but are not limited to:
pyroxene (augite) *or* mica (biotite) *or* amphibole (hornblende)

Explanation: Open to the Metamorphic Rock chart and locate gneiss. Move to the Composition column. Here it shows (by the gray mineral bars) that all six given minerals are found in gneiss. Locate these six minerals in the Properties of Common Minerals chart. Check the Composition column for these minerals to see if they contain Fe (iron) or Mg (magnesium).

64. Mineral name: Garnet
Acceptable use: jewelry *or* abrasives

Explanation: Open to the Properties of Common Minerals chart and locate garnet. The given color and luster match that of the given mineral. In the Use(s) column, it shows it is used for jewelry and abrasives.

65. Answer: July 7 *or* July 8

Explanation: When we see a full moon, we are seeing all (100%) of the reflected light of the moon. The moon phase in this question is a quarter moon in which we see 50% of the reflected light. The chart shows this would have occurred between July 7 and July 8.

66. Acceptable responses include:
The Moon's revolution.
or The Moon orbits Earth.

Explanation: The Moon takes about one month to orbit the Earth. As it revolves around the Earth, the angle of the Sun - Moon - Earth changes, causing us to see different amounts of the lighted side of the Moon. This produces the phases of the Moon.

67. Answer: August 12 *or* 13 *or* 14

Explanation: The time period from a full moon to the next full moon is just under one month. The next time you observe a full moon, make a note of the date and see if it takes just about another month to see the next full moon.

68. 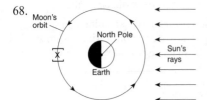 **X** must be placed in the bracket area.

Explanation: A full moon occurs when all the reflected light (100%) of the Moon is seen from the Earth. This occurs when the Moon is positioned on the opposite side of the Earth with respect to the Sun.

69. Acceptable responses include, but are not limited to:
The Moon phases repeat in a definite pattern.
or The visible part of the Moon increases and decreases repeatedly.
or The Moon phases are predictable.

Explanation: A cyclic event is one that repeats in a predictable manner. All cyclic events can be predicted into the future as with the moon phases. That cyclic predictability makes it possible to put the moon phases on calendars.

70. Answer: 10 m

Explanation: Position A is in a depression, a place where the elevation values go down. The contour interval is 10 meters. This means that every contour line goes up or down by this value. Position A is one line down from the 20 meter depression contour line. This makes the elevation of A 10 meters.

71. Answer: Any value from 18.9 m/km to 21.1 m/km

Explanation: The gradient equation is: $G = \dfrac{\text{change over field value}}{\text{distance}}$

The elevation of B is 10 m, while the elevation of C is 50 m. This gives us a change in field value of 40 m. Using a piece of paper, mark off the distance between B and C. Using this distance and the given scale below the map, the distance measures 2 km.

Solution: $G = 40 \text{ m}/2 \text{ km} = 20 \text{ m/km}$.

72.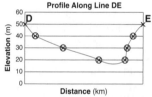

Explanation: Place a piece of paper along line D-E on the topographic map. Make a mark on the edge of this paper where each contour line intersects the paper. Place the edge of this paper with the marks along the bottom line of the profile graph. Move vertically up the graph above each mark to its correct elevation, and make an **X** at the position. Connect the **X**'s making sure the profile line drops slightly below the lowest point (20 m) and the profile line extends completely from point D to point E.

73. Acceptable responses include, but are not limited to:
Contour lines between N and M are closer together.
or There is a steeper slope between N and M
or Where contour lines are far apart, there is a gentle slope and the stream velocity is less.

Explanation: Whenever isolines of any type are close together, it indicates a steep gradient. When this is observed on a topographic map, it indicates that the elevation is changing quickly. This reflects the situation between point N and point M, and in this steep gradient area the stream velocity (speed) would increase. Between points K and L there are no contour lines crossing this area. Thus there is no change in elevation, making this area flat.

June 2007 Answer Keys

74. Answer: From 6.0 to 6.2

Explanation: Reread the example given on the previous page to understand the procedure to determine the Richter magnitude of an earthquake. Using a straight edge (like a piece of paper) measure the height of highest S-wave above the baseline. If done correctly, it is 20 mm. Mark this value on the Height of the largest S-wave scale. The distance to the epicenter is given as 500 km. Mark this position on the Distance to epicenter scale. Placing a straight edge from the 20 mm mark to the 500 km mark, draw a straight line between these points (see Example page). If done correctly, the line crosses the Richter magnitude scale very close to the 6.1 value.

75. Acceptable responses include, but are not limited to:
the lag time between the *P*-wave arrival and the *S*-wave arrival
or the difference in arrival time for the *P*-wave and *S*-wave
or the *P*-wave and *S*-wave arrival times
or 61 seconds

Explanation: To determine the distance to the epicenter, one must know the separation time (which is the lag time) between the *P* and *S*-wave arrival time. This separation time of 61 seconds, is marked off on a piece of scrap paper using the Travel Time axis located on the Earthquake *P*-wave and *S*-wave Travel Time chart. When this distance is fitted vertically between the *P* and *S* lines, the epicenter distance is read directly down on the *x*-axis.

76. Answer: From 2 minutes 0 seconds to 2 minutes 20 seconds.

Explanation: Open to the Earthquake *P*-wave and *S*-wave Travel Time graph. Locate the 500 km position on the Epicenter Distance axis, knowing that every line equals 200 km. From this distance, move directly upward until it intersects the *S*-wave line. At this intersection point, read the time on the Travel Time axis. If done correctly, 2:00 to 2:20 is the time of the arrival time of the first *S*-wave.

77. 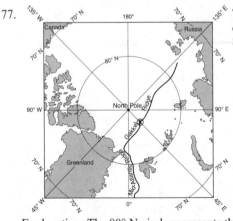 The letter **X** indicates the correct coordinates of 87° N, 45° E.

Explanation: The 80° N circle represents the 80° N latitude line. The dot at the North Pole is the 90° N latitude position – the highest latitude measurement. The hydrothermal vent is on the Gakkel Ridge (87° N latitude line) and at the intersection of the 45° E longitude line. Remember, all longitude lines meet at the Poles and are measured E and W of the Prime Meridian.

78. Acceptable responses include, but are not limited to:
The plates are moving apart or spreading.
or The tectonic plates are moving away from each other.
or The ridge is a diverging plate boundary.
or rifting

Explanation: The reading states that the Gakkel Ridge is a section of the Arctic Mid-Ocean Ridge, which is a divergent plate boundary. The motion of divergent boundaries is a spreading motion, moving away from the ridge. This plate boundary and its motion is diagrammed in the Key area of the Tectonic Plate map in the RT .

79. Answer: North American Plate and Eurasian Plate

Explanation: Open to the Tectonic Plates map in the RT and locate the Mid-Atlantic Ridge that cuts through Iceland. As the article explains, this ridge continues north of Iceland into the Arctic Ocean. The North America Plate is to the left and the Eurasian Plate is to the right of this ridge.

80. Acceptable responses include, but are not limited to:
magma/lava *or* volcanoes *or* smoker vents *or* hotspots

Explanation: At mid-ocean ridges, magma moves upward and escapes as lava through openings located in ridge areas. At times the lava builds upward, creating volcanoes as found on Iceland. Hotspots are volcanic active areas that may be found on ridges. Another interesting feature found along ridges are bellowing black sulfur smokers. These poisonous gases have there origin deep within the interior of the Earth.

81. Answer: pyroxene (augite) or olivine

Explanation: Open to the Igneous Rock chart in the RT. Locate the rock Peridotite on the right side. Directly down from the peridotite position in the Mineral Composition chart are the minerals that would be found in peridotite. This chart only shows two minerals that make up this igneous rock, being pyroxene and olivine.

82.

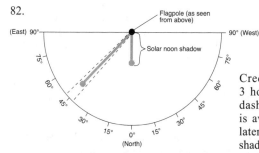

Credit is awarded if the shadow, 3 hours later, is drawn within the dashed lines shown above. Credit is awarded if the shadow, 3 hours later, is longer than the solar noon shadow.

Explanation: At solar noon in NYS, the Sun is at its daily highest position and located in the southern part of the sky. This produces the shortest daily shadows pointing to the north (a shadow is always opposite the direction of the Sun). Three hours later, the Sun has appeared to have moved 45° to the west due to the motion of rotation. This causes the shadow to move 45° to the east. Since the Sun altitude decreases after solar noon, the lower the Sun gets, the longer the shadow becomes.

June 2008

1. 2   Rotation is the spinning motion around an axis. Our day is based on the Earth's rotation. It is this motion that causes celestial objects (Sun, stars, moon, etc.) to appear to move daily across the sky.

2. 1   Open to the Luminosity and Temperature of Stars chart in the reference tables. On this chart Aldebaran has a luminosity of just over 100, while Rigel is shown to have a temperature close to $12,000°C$. The intersection position of these values falls within the Main Sequence, therefore Algol would be classified as a main sequence star.

3. 4   Scientists have concluded that the formation of our Universe by the Big Bang theory occurred over 10 billion years ago. This was inferred by a number of methods, such as measuring the red shift of spectral lines from very distance celestial objects and by measuring the temperature and amount of left over radiation from the Big Bang. The 4.6 billion years answer is the estimated time for the origin of the Earth and solar system.

4. 2   Because of revolution, the Earth moves in its orbit throughout the year. This motion positions the Earth to see different constellations at different times throughout the year. If a seasonal constellation is visible at night, six months later the Earth will be positioned on the other side of its orbit. At this time, this constellation will be positioned in the day sky and will not be visible being outshined by the Sun.

5. 1   The Coriolis effect is the deflection or curvature of moving objects that travel over the surface of our rotating planet. Moving air masses and ocean currents both experience the Coriolis effect, changing their direction as they travel over our rotating planet. The Coriolis effect is an accepted proof of rotation.

6. 2   Open to the Selected Properties of Earth's Atmosphere chart in the reference tables. On the Altitude scale locate 20 km (not miles). At this height, the chart shows that the temperature zone is the stratosphere. This layer contains most of the ozone.

7. 3   Open to the Weather Map Symbols chart in the reference tables. In the Air Masses section five different air masses are given, each represented by two letters. One letter represents the relative temperature of the air mass and the other letter stands for the relative moisture content of the air mass.

8. 4   In the Northern Hemisphere, low-pressure systems have a surface air motion of counterclockwise and inward to the center. High-pressure systems have a surface air motion of clockwise and out from the center.

9. 3   Open to the Specific Heats of Common Materials chart in the reference tables. The lower the specific heat value is, the faster the substance heats up and cools down compared to water. Land is represented by the Earth's materials of granite and basalt. These two substances, having low specific heat values, heat up and cool down about five times faster than water. Thus when winter arrives, land quickly gives up its heat and the ground freezes. Water, having a high specific heat value, cools down much slower and remains in the liquid phase much longer. The Finger Lakes, being so large, usually remain unfrozen throughout the winter.

10. 1 Nuclear fusion is the combining of lightweight nuclei to produce a heavier nucleus. Within our Sun (stars), almost of all the ongoing fusion involves the joining of hydrogen nuclei to producing one heavier helium nucleus, as the given equation shows. In this nuclear reaction, a small amount of mass is converted to a large amount of energy. This reaction occurs at a tremendous rate within our Sun and the released energy we receive as radiant energy.

11. 2 As the Moon revolves around the Earth, different amounts of light are seen being reflected from the Moon. At position *A*, the moon phase will be waxing crescent. As the Moon revolves from position *A* to position *B*, the Moon's light increases (waxing), as seen from the night side of the Earth. At position *B*, waxing gibbous phase is occurring and it would look like Diagram 2. In about 3 more days the Moon would be positioned between *B* and *C* and full moon phase would occur.

12. 1 In the Northern Hemisphere, the angle of Polaris above the northern horizon is equal to the latitude of the observer. This observer must be positioned on the 20° N latitude line. Remember as one travels north, the angle of Polaris increases up to 90°.

13. 2 Spectral lines are bright color lines that are emitted (given off) by elements at high temperatures. If a radiating object is moving away from the Earth, the wavelengths of the spectral lines become longer and move toward the red end of the spectrum. This is known as the red shift.

14. 2 The dry-bulb temperature (air temperature) is 18°C, and the wet-bulb temperature is 10°C, a difference of 8°C. At the top of the Dewpoint Temperatures chart, locate the Difference of 8°C column. In this column move down until you reach the Dry-Bulb Temperature row of 18°C. The intersection of these two readings gives a dewpoint temperature of 2°C.

15. 3 Tides are caused by the gravitational pull of the Moon and Sun. The Moon is responsible for most of the tidal action since it is closer to the Earth than the Sun. In choice 3, the Sun and Moon are aligned and the combined gravitational forces produce the greatest bulge of the ocean water. This causes the highest high tides and at the same time results in the lowest low tides. These are known as spring tides, though they have nothing to due with the season.

16. 2 Refer to the Weather Map Symbols – Station Model chart in the reference tables. Choices 1 and 2 both show the wind coming from the NE at 25 knots. The change of the barometric pressure over the past three hours is known as the barometric trend. As shown on the Station Model, this value is always placed in the middle of the right side of the station model.

17. 3 Open to the Surface Ocean Currents chart in the reference tables. The key shows the two classifications of these currents, warm currents and cool currents. The Peru Current is a cool current moving north up the western side of South America toward the equator.

18. 4   Instrument *A* is a barometer showing pressure values in inches of mercury (see the Pressure chart in the reference tables). This instrument may have weather related words on it, since the lower the pressure the more likely stormy weather is occurring. Instrument *B* is an anemometer, which measures wind speed. The cups are spun by the wind and the instrument's needle shows the calibrated speed of the wind.

19. 4   Rough, dark surfaces absorb the most sunlight (insolation), while white, smooth surfaces reflect the most insolation.

20. 3   Corals live in tropical warm shallow seas as shown on the given map. Millions of years ago during the Devonian Period, North America was near the equator and coral communities were abundant in the seas that existed at that time in and around NYS. Due to plate tectonics, North America slowly drifted northwest away from the equator. Due to the colder climate corals died off. Their fossil remains can be found in some of the exposed Devonian-age rocks located in NYS. The changing position of continents over geologic time is diagrammed in the Geologic History of NYS – Inferred Position of Earth's Landmasses column found in the reference tables.

21. 3   Open to the Generalized Bedrock Geology of NYS map in the reference table. The surface bedrock of Ithaca is Devonian in age and would contain Devonian fossils. Go to the Geologic History of NYS chart and in the Time Distribution of Fossils column locate letter R on the Placoderm Fish bar. As diagrammed on the top right of this chart, letter R represents the fossil fish Bothriolepis. Since letter R is positioned in the Devonian row, it lived during this period. All other shown fossils are from other geologic time periods.

22. 2   For a fossil to be an index fossil, it must have the characteristic of being widespread (lived in many locations) and lived a relatively short time span, being found in one geologic rock layer. Fossil B is widespread, located in all three exposed outcrops, and lived a relatively short time span, found in only one layer (the sandstone layer). This fossil would make a good index fossil.

23. 1   Open to the Tectonic Plates map in the reference table. Here it shows the location of plate boundaries where plates are diverging, converging or slipping (transform boundaries) by each other. Along these relatively thin, cracked plate boundaries, magma rises from the asthenosphere surfaces to form volcanic features. Other areas of the Earth have active volcanoes (Hot Spots) but by far, more active volcanoes are located on the edges of tectonic plates.

24. 2   Open to the Inferred Properties of the Earth's Interior chart in the reference table. The arrows shown in the asthenosphere represent convection currents. These very large, continuously flowing currents are the major force that slowly move plates to new positions.

25. 4   Open to the Inferred Properties of the Earth's Interior chart in the reference table. On the upper right side of the cross-section, the densities of the plates are given. The ocean plate is denser than the continental plate. The composition of the ocean plate is basaltic, while the continental plate is granitic. Open to the Igneous Rock chart and notice that basalt is located on the right side of this chart. Move directly down from basalt to the Characteristics section – Composition line. For rocks on the right side (where basalt is located), the composition shown is MAFIC. These rocks contain a high percentage of Fe and Mg.

26. 1 The graph shows that element X is the second most abundant element in the Earth's crust by mass. Open to the Average Chemical Composition of Earth's Crust, Hydrosphere, and Troposphere chart. Here it shows that within the crust, silicon (Si) is the second most abundant element by mass and by volume.

27. 1 Landscape regions A and B have similar sedimentary rock layers while being close to the same elevation. Their different landscape features have been formed by being exposed to different climates over a long time period. In a humid climate (as in NYS), the landscape features take on a rounder appearance as diagrammed in A. In an arid climate (as in Arizona – location of the Grand Canyon), landscapes tend to have more cliffs or straighter wall features as diagrammed in B.

28. 3 The faulting has caused major displacement of the rock layers. This, along with the interaction of weathering and erosion over much time, produced deep valleys. On the floor of each valley will be the location of a separate river. Side streams, known as tributaries, will flow down the valley walls into the main valley river. When this landscape is viewed from a high altitude, the observed stream pattern will be similar to choice 3.

29. 2 The discharge of a stream refers to the amount of water a stream contains. After a heavy rain, the discharge of a steam increases. As more water enters a stream, the speed or velocity of a stream also increases. This direct relationship is shown by graph 2.

30. 4 In a stream when water rounds a curve, the streambed at the inside of curve (A) experiences deposition of sediments, while the streambed on the outside of curve (C) undergoes erosion. The middle section of the streambed (B) will be in a dynamic equilibrium state, in which some sediments are deposited as other sediments are eroded. The net results of these actions result in a situation that does not appear to experience any change.

31. 3 Mass movement is the process by which large amounts of sediment (or at times snow) moves downhill under the force of gravity. Steep slopes with saturated soil greatly contribute to mass movements. Earth scientists have identified different types of mass movements such as rockslides, mudflows and creep. Mass movements (wasting) may occur at a very slow rate (soil creep), or at high speeds (mudflows), which can cause much destruction and loss of life.

32. 4 Deposition of sediments occur when a stream's velocity (speed) decreases. The major reason a stream undergoes deposition is due to a decrease in the gradient of the land which causes a stream's velocity to slow up. This is occurring at point D as the land becomes more level. As the velocity of the water decreases, more sediment will be deposited. This relationship is represented in the Relationship of Transported Particle Size to Water Velocity graph in the reference table.

33. 2 Open to the Relationship of Transported Particle Size to Water Velocity chart in the reference tables. At the 100 cm/sec position, move directly up to the intersection of the graph line. This occurs in the middle of the Pebbles section. A stream with this velocity can transport medium size pebbles and all the listed smaller sediments under the graph line. Thus, as the stream slows down from 250 cm/sec to 100 cm/sec, it will first deposit the larger sediments, which are cobbles, and then some of the larger pebbles.

34. 4   Open to the Sedimentary Rock chart in the reference table. Dolostone is located in the chart titled Chemically and/or Organically Formed Sedimentary Rocks. In the Comments section it gives information that classifies dolostone as an evaporate. These rocks are formed as water evaporates, concentrating the minerals (chemicals), which precipitate out of the water to form these rocks.

35. 4   Open to the Igneous Rock chart and in the Texture section locate the box labeled Vesicular (gas pockets). This matches the photograph of the given rock. On the left side, directly across from Vesicular, it gives the Environment of Formation as Extrusive. As lava is ejected it quickly cools, trapping air, and forming gas pockets. This vesicular texture is an excellent clue for igneous extrusive rocks that experienced rapid cooling.

Part B-1

36. 3   As shown in the diagram, when Asteroid Hermes is at position *A*, it is on the night side of the Earth. The asteroid would be visible as the sky darkens at sunset and would remain visible until sunrise. At noon, an observer would be facing the Sun and Asteroid Hermes, being on the opposite side of the Earth, would be out of view.

37. 3   The farther a celestial object is from the Sun, the longer its period of revolution will be. Asteroid Hermes has an orbit that extends farther out than all of the shown planets. Thus, it would have a longer period of revolution than the shown planets.

38. 4   Due to the existence of our atmosphere and water, our Earth is constantly being weathered and eroded. Over tens of thousands of years, these actions would erase evidence of an impact crater. Arizona is the home of one of the most famous impact crater - Meteorite Crater. It is slowly being worn away by weathering and erosion, while slowly being covered by the deposition of wind blown sediments.

39. 3   The orbit of Asteroid Hermes passes through the orbits of all the shown planets except for Mercury. Thus Mercury will never be in danger of colliding with Hermes. Not so for the other three planets. As they revolve around the Sun, one day in the future, they could be in a position to have an impact with this asteroid.

40. 4   Open to the Generalized Landscape Regions of NYS map located in the reference tables. Matching exposed Grenville rocks with the landscape regions of NYS, the largest exposure is found in the Adirondack Mountains and smaller amounts are located in the Hudson Highlands.

41. 4   City A has a temperature range of 25°C to 30°C. Open to the Temperature scale in the reference table and it shows that this temperature range is close to 75°F to 85°F. This warm temperature range, having little variation, would be expected near the equator due to the yearly strong (direct) sunlight.

42. 2   For City *B*, July has the greatest precipitation amount as shown by the height of the gray bar for this month. This is during our summer.

43. 3 The seasons are reversed in the Southern Hemisphere relative to the Northern Hemisphere. When we are having summer, they are having winter. From the graph of City *C*, it is experiencing the lowest temperatures during July and August (being its winter) and the warmest temperatures during January and February (being its summer).

44. 1 Water infiltrates (enters) the soil when the soil is permeable. When the ground is frozen, it becomes impermeable and water cannot infiltrate. City *D*'s low temperatures would make for a frozen surface year round, which would prevent water from entering.

45. 4 Layers *A*, *B*, *C* and *D* are sedimentary rock layers that were formed horizontally underwater. Later in time these layers were uplifted to their present tilted position. Sometime after this, all of these rock layers were intruded or cut through by the igneous intrusion as evidence by contact metamorphism. This makes the intrusion the youngest rock layer. Over time, the intrusion solidified and is now being attacked by agents of erosion.

46. 1 Open to the Igneous Rock Identification chart found in the reference table. Locate the Coarse Texture row having a grain size of 1 mm to 10 mm (where 3 mm grain size would fall). In this row, move to granite. Directly under granite in the Mineral Composition chart, potassium feldspar and quartz are listed. These two minerals must be in granite along with plagioclase feldspar, biotite, and amphibole.

47. 4 Open up to the Sedimentary Rock Identification charts in the reference table. From the Map Symbols, layer *B* is the sedimentary rock – sandstone. In the Rock Cycle chart, it shows that compaction and cementation are processes that form sedimentary rocks.

48. 3 Moraines are large hills consisting of unsorted glacial rocks. Where a glacier undergoes melting, the suspended rocks fall out and accumulate upon each other, eventually producing a large hill called a moraine.

49. 3 Position *E* is near the Harbor Hill Moraine. Moving toward position *F*, one would cross over this moraine into the outwash plain produced by the melt water of the glacier. Continuing south, one crosses the Ronkonkoma Moraine and enters the other outwash plain. The order of these features matches the ones shown in the cross section diagram.

50. 2 Moraines are glacial deposition features formed by the deposition of large amounts of unsorted sediments where the glacier is melting. In an outwash plain, the sediments are transported by the melting glacial water. Water deposits sediments by their sizes, with the larger ones being deposited first and the lightest last. This causes a sorting pattern of sediments.

Part B-2 and C

51. Answer: The drawn line should be at the same level as the black line shown in the diagram below.

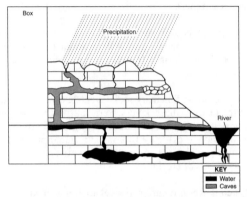

Explanation: The water table is the top of the subsurface water (underground water). Within this cave, the water table is equal to the river level. Overtime, the height of the water table will change due to the amount of precipitation the area receives.

52. Answers include: The acid rain dissolves the limestone.
or The calcite in the limestone chemically reacts with the acid.

Explanation: In the reference tables, open to the Sedimentary Rock Identification chart and locate limestone. In the Composition section it states that limestone is composed of the mineral calcite. Open to the Mineral Identification chart and locate calcite. In the Distinguishing Characteristic section of calcite it states "bubbles with acid." As rain becomes more acidic, calcite will weather faster.

53. Answers include: burning fossil fuels
or exhaust emissions from automobiles
or smoke from factories

Explanation: All of the above answers are sources for the formation of nitric acid and sulfuric acid when combustion occurs. When airborne, these acids chemically join with moisture producing acid rain.

54. Answer: Mercury

Explanation: The given diameter of Callisto is 4,800 km. In the Solar System Data chart, Mercury is shown to have a diameter of 4,880 km, and their densities are relatively close.

55. Answers include: These moons orbit Jupiter, not Earth.
or The geocentric model has all celestial objects revolving around Earth.

Explanation: In the geocentric model, all celestial objects revolved around the Earth. Galileo observed these moons revolving around Jupiter, which contradicted the geocentric model. Using this and other proofs, Galileo eventually rejected the geocentric theory and gave support to the heliocentric theory. In this theory, the Sun is the center of the solar system.

56. Answer: See diagram below. The warm front symbol must be drawn on the correct side of line *XY*.

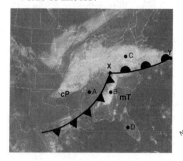

Explanation: Line *XY* represents a warm front that moved northeast from the Gulf of Mexico. The symbol associated with warm fronts is found in the reference table -Weather Map Symbols chart. The warm front symbol is drawn on the side of the frontal line that it is moving toward. Remember, the warm air is always behind the warm front.

57. Answers include: condensation *or* expanding air *or* cooling to the dewpoint *or* rising air *or* deposition (sublimation)

Explanation: Along the cold front, the cold, denser air is forcing the moist, warmer, less dense air upward. This rising air will expand causing a cooling effect. As the air's temperature drops, it will eventually reach the dewpoint temperature and condensation of the water vapor will occur.

58. Answers include: Location *A* is influenced by a cold, dry air mass. *or* Location *A* has clear skies. *or* Location *B* is in a warm, moist air mass. *or* Location *B* has cloud cover.

Explanation: Within a cloud the relative humidity is 100%, thus saturated with moisture. This occurs along frontal boundaries where air is rising and the dewpoint temperature has been reached, causing condensation. Location *B* is cloudy, therefore will have a high relative humidity. At location *A* the the cold front has already passed causing the air to be colder and drier. These conditions produce low relative humidity resulting in few or no clouds.

59. Answers include: Location *C* is cooler because it is farther north. *or* *C* is in a continental polar air mass, which is cold, dry air. *or* Location *C* has clouds that block some of the sunlight (insolation).

Explanation: Generally the further north one travels, the colder it becomes since the Sun's rays becomes more indirect. From the diagram, the warm front has not reached Location *C*, so the cP air mass is still influencing the weather at Location *C*. cP air masses have the characteristic properties of being cold and dry.

60. Answers include: east *or* northeast *or* ENE

Explanation: The center of this low-pressure system is near location *X*. On a weather map, this position would be labeled with an "L". Most low-pressure systems move towards the northeast as they travel across the United States.

61. Answers include: *P*-waves can travel through the liquid outer core, but *S*-waves cannot. *or* *P*-waves travel through all parts of Earth's interior. *or* *S*-waves do not pass through the outer core.

Explanation: Seismic station *A* is almost on the other side of the Earth with relationship to the location of the epicenter. As the seismic waves travel towards *A*, they must pass through the outer core. This core, being a liquid, stops the *S*-waves. The *P*-waves can be transmitted through this and all other layers.

62. Answer: 2800 to 3000 km.

Explanation: Location *B* is on the boundary line of the mantle and outer core. This depth is given in the Inferred Properties of Earth's Interior chart found in the reference table. In this chart, locate this position and move downward to the dashed line. Follow this dashed line until it reaches the Depth axis. It intersects the Depth axis at approximately 2900 km.

63. Answer: Any value from 1408 to 1409 millions of cubic kilometers.

Explanation: The chart shows that the total volume of water stored in the atmosphere is 0.013 millions of cubic kilometers. The oceans stores 1370.000 million cubic kilometers, and the continents holds 38.631 million cubic kilometers. Adding these amounts totals 1408.6 million cubic kilometers of stored water.

64. Answers include: The oceans cover a larger portion of Earth's surface than the continents.
or Air over oceans has more moisture than air over land.
or More evaporation occurs over the oceans.

Explanation: By far, the oceans are the major source of water for our planet. Over the oceans much evaporation occurs daily. Thus, it would be expected to have more precipitation over the oceans compared to land surfaces.

65. Answers include: slope of land surface
or soil type or composition *or* vegetation *or* lack of vegetation
or a paved surface (like blacktop or concrete) *or* how saturated the soil is
or porosity (air spaces) of the soil
or permeability of the surface *or* impermeable of the surface

Explanation: Precipitation will either infiltrate (enter) the ground, evaporate back into the atmosphere or it may become surface runoff and enter streams flowing towards the oceans. If the ground is impermeable (not allowing water to enter), the water becomes runoff. The answers given above are the major reasons that will affect the rate of stream runoff.
Note: Clay and solid bedrock are impermeable, which will cause much runoff.

66. Answer: See the graph below. All points must fall within the circles located on the graph shown below.

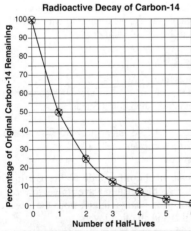

Radioactive Decay of Carbon-14

Explanation: On the Half-Lives axis, locate the correct number of half-lives and move directly upward until it intersects the correct line for Percentage of Original C-14 Remaining. Place an **X** at this intersection point. Continue this process for all given half-lives. Connect the **X**'s with a smooth curved line.

67. Answer: 22,800 yr.

Explanation: The half-life of C-14 is 5,700 years. This is shown on the given chart and in the Radioactive Decay Data chart in the reference table. At the completion of the 3rd half-life of C-14, 17,100 years have passed. To obtain the age at the completion of the 4th half-life, add 5,700 to 17,100 for a total of 22,800 years.

68. Answers include: The tree trunk is a recent organic remain.
or Carbon-14 is used to date recent remains.

Explanation: Carbon 14 has a half-life of 5,700 years. When compared to the geologic time line, this half-life is relatively short. Therefore C-14 is used to date organic material that is relatively young (usually less than 60,000 years old). The shown tree trunk is buried in Pleistocene glacial deposits. Open to the Geologic History of NYS chart found in the reference table and locate the Pleistocene Epoch. Being near the top of the geologic time scale time period, this epoch is relatively young. C-14 could be used to date a tree trunk from this age.

69. Answer: chromium *or* Cr

Explanation: This is stated in the 3rd sentence of the passage.

70. Answers include: hardness *or* luster *or* crystal shape

Explanation: Gemstones usually have a hardness of 7 or higher on the Mohs hardness scale. These semiprecious or precious gemstones usually have a desirable crystal shape and produce a brilliant luster.

71. Answer: marble

Explanation: The passage states "The limestone rock lining the seafloor underwent metamorphism…" . Metamorphism of limestone produces marble as shown in the Metamorphic Rock Identification chart (see Comments section for Marble).

72. Answer: Eocene Epoch.

Explanation: The passage states that the ruby deposits occurred around 50 million years ago. Open to the Geologic History of NYS chart and find 50 mya on the time line just to the right of the Epoch column. Notice that the Eocene Epoch spans 54.8 mya to 33.7 mya.

73. Answers include: convergent plate boundary *or* subduction zone
or collision boundary

Explanation: The cross section shows two plates moving toward each other. These are convergent plates which will produce a subduction zone. This plate boundary and the other boundaries are shown at the bottom of the Tectonic Plates map.

74. Answers include: The Sun will appear higher in the sky as Earth moves to position *B*, then lower in the sky as it moves to position *C*.
or The angle increases as it approaches *B*, and decreases as it approaches *C*.
or Altitude will increase, then decrease.

Explanation: In Diagram 1 - Location A, the Northern Hemisphere is experiencing spring, at Location *B* the Northern Hemisphere is experiencing summer and at Location *C* it is fall. The Sun's noon altitude increases each day as the Earth travels from its spring position to its summer position. On the first day of summer, the Sun will have its highest noon altitude. For the next 6 months, the Sun's noon altitude decreases. This change of the Sun's noon altitude is due to the Earth's tilt on its axis and motion of revolution.

75. Answers include: On an equinox, all locations will have the same duration of insolation.

   or The Sun's direct rays are at the equator on this day.

   or Each location is in sunlight for half of the 24-hour Earth rotation.

   Explanation: Positions C and A are the equinoxes, being the first day of fall and spring, respectfully. At this time, the Sun's rays are direct (90°) on the equator at noon and all positions on the Earth, including both Poles, and receive 12 hours of sunlight. This night and day equal division for the whole Earth can be seen in diagrams C and A.

76. Answers include: Earth's Northern Hemisphere is tilted away from the Sun. During the winter the Northern Hemisphere experiences less hours of sunlight compared to the other seasons.

   or The Arctic Circle is dark and has 24 hours of night.

   or Less sunlight is received in the Northern Hemisphere.

   Explanation: Position D shows that the Artic Circle is totally in the dark and would stay in darkness 24 hours as the Earth rotates around the axis. This occurs in the winter because the Northern Hemisphere is tilted away from the Sun, producing less hours of sunlight compared to all other seasons.

77. Answer: 4 p.m.   *or*   1600 hours

   Explanation: Diagram 2 is a polar view of the Earth. Each 15° segment, shown by longitude lines, represents a time interval of one hour and all 24 hours (segments) are drawn. On March 21 the Earth is at equinox and is the first day of spring. At the equinoxes, sunrise occurs at 6 a.m. and sunset occurs at 6 p.m. Position E is two hours from sunset so its time must be 4 p.m.

78. Answer: See map below.

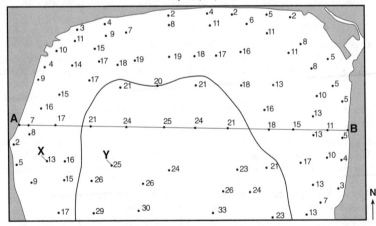

Water Depths (feet)

Explanation: Starting from the edge of the map, estimate where the 20-foot-depth position would be by using the given depths. Continue to estimate the 20-foot-depth positions and extend your line through these estimated points, working toward the edge of the map. For the given exact 20-foot mark, make sure your isoline passes through this position. The completed isoline must extend to the edge of the map to receive credit.

79. Answer: See profile below.

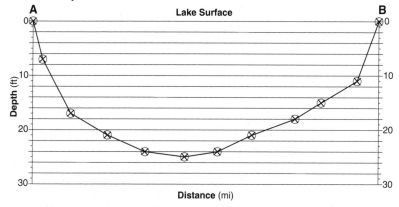

Explanation: Place a piece of paper along line *A-B* on the map. Make a mark on the edge of this paper where each numbered point representing a water depth to its correct elevation. Place the edge of this paper, with the water depths marks, along the bottom line of the given profile graph found in your answer booklet. Move vertically up the graph above each water depth to its correct elevation. Place an **X** at each correct elevation. Connect the **X**'s with a smooth curved line. All of your **X**'s must fall within the circles shown in the above answer key.

80. Answer: 30 feet/mile ± 2.0

Explanation: Explanation: The gradient equation is: $G = \dfrac{\text{change over field value}}{\text{distance}}$.

The elevation of *X* is 13 ft, and the elevation of *Y* is 25 ft. This gives a change in field value of 12 ft. With a piece of paper, mark off the distance between *X* and *Y*. Using this distance and the given scale on the map, the distance measures 0.4 miles. Substituting: G = 12ft/0.4 miles = 30 ft/mile

81. Answer: Latitude: 34° N Longitude: 118.5° W (118°30′W)

Explanation: Using a straight edge, the latitude is very close to 34° N. The longitude is very close to 118.5° W or 118°30′W. Latitude and longitude degrees can be subdivided into 60 minutes (60′), thus 0.5° is equal to 30′. Remember, for a coordinate reading, latitude goes first, then longitude.

82. Answers include: It is near a plate boundary.
or The San Andreas Fault is nearby.
or The bedrock contains many faults.

Explanation: Open to the Tectonic Plates map in the reference table. Locate the San Andreas Fault line in California. Notice that this fault line is a transform plate boundary between the Pacific Ocean Plate and the North American Plate. As plates shift, earthquakes are generated. Thus, most earthquakes, and almost all major ones, have their epicenter along plate boundaries, of which this region of California is located on one.

83. Answer: Oakland is farthest from the epicenter.

Explanation: Northridge was the epicenter of this earthquake. Seismic waves radiated outward from this position causing damage along the way. The farther a city is from an epicenter, the longer it takes for the different seismic waves to reach it.

84. Answers include: secure heavy objects
    or prepare an emergency medical kit
    or plan an evacuation route
    or locate the nearest shelter
    or reinforce house structure.
    Note: other answers are possible.

Explanation: Much research and engineering has been devoted to making structures "earthquake proof." Many of these design improvements have been implemented in building structures. As a homeowner, other actions can be taken within the home to prepare for earthquakes. Some of these actions are given above. There are many other acceptable actions that improve home safety in the event of an earthquake.

Relationships

Within Earth Science we find certain relationships between two variables that are always constant and thus can be represented by one of four graphs: a direct or proportional graph; an inverse or indirect graph; a cyclic graph and a graph that remains the same. In a direct or proportional graph both variables are increasing and the resulting line will always slope up; in a inverse graph one variable is increasing while the other is decreasing producing a graph with the line sloping down; in a cyclic graph one variable is repeating and the line shows a repeating / predictable pattern; and if one of the two variables is not changing we get a graph with a straight horizontal line producing a graph representing no change.

Graph examples

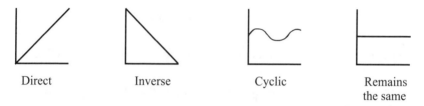

Direct   Inverse   Cyclic   Remains
                                                  the same

Examples of some **direct relationships** in Earth Science

Pollution – As the population increases, so does the amount of pollution

Permeability – As the grain size increases, so does the permeability.

Gravity – As the mass of two objects increase, so will the gravitational attraction increase.

Evaporation – As the temperature goes up, so will the evaporation rate.

Polaris – As your latitude increases, the altitude of Polaris increases.

Chemical weathering – As the temperature goes up, the rate of chemical weathering increases.

Coriolis effect – As the speed of rotation of a planet increases, so will the Coriolis effect.

Pressure within the Earth – The deeper within the Earth, the greater the pressure.

Density within the Earth – The deeper within the Earth, the greater the density.

Streams – The steeper the gradient, the faster the stream's velocity.

Runoff – The greater the runoff, the faster the stream's velocity will be.

Greenhouse effect – The more greenhouse gases (carbon dioxide, water vapor, methane, etc.,), the warmer the atmosphere will become.

Pressure gradient – As the pressure gradient increases (shown by isobars), the faster the wind velocity.

Mass and Volume – As the mass of an object increases, so will its volume.

Seismic waves – The further the seismic waves (P and S) travel, the greater the separation time will be.

Angle of insolation – As the angle of insolation becomes more direct, the heating power increases.

Water table – Under normal conditions, as rainfall increases, so will the water table's level.

Soil development – In the absence of erosion, as time increases, soil horizons (layers) increase.

Mid Atlantic Ridge – The further you go from the MAR, the older igneous rocks become.

Eccentricity – As the foci distance increases, the more eccentric the planet's orbit will become.

Eccentricity – As the eccentricity number increases, the orbit becomes more elliptical.

Red Shift –The further a galaxy is from us, the greater the red shift.

Deposition – The denser the sediment particles are, the faster it settles.

Examples of some **inverse relationships** within Earth Science

Revolution period – As the distance to the sun decreases, the orbital speed will increase.

Density change – As the temperature of the air increases, its density decreases.

Water Vapor – As your altitude increases in the atmosphere, the amount of water vapor decreases.

Water vapor – As more water vapor is added to the atmosphere, the lower the air pressure will be.

Air pressure – As your altitude increases in the atmosphere, the pressure decreases.

Radioactivity – As time goes on, the radioactivity decreases within a sample.

Deposition – As the speed of a stream decreases, the deposition of sediments increases.

Specific Heat – The higher the specific heat number is, the slower the substance heats up.

Hurricanes – As a hurricane strengthens, the air pressure decreases.

Physical weathering – As temperature drops, physical weathering increases (frost action).

Atmospheric transparency – As air pollution increases, the atmospheric transparency decreases.

Air temperature – As air expands, the temperature decreases.

Soil capillarity – As the size of the soil particles decrease, soil capillarity increases.

Infiltration – As the slope increases, the infiltration decreases.

Examples of **cyclic changes** in Earth Science

Plotting the following would produce a cyclic graph;

Moon phases

Sunspots

Tides

Seasons

Sun's path

Examples of **remaining the same graphs** in Earth Science

Half-life – As time goes by, the half-life of an element remains the same.

Latitude – As one travels along a latitude line the altitude of Polaris stays the same.

Longitude – As one travels along a longitude line the local solar time remains the same.

Apparent daily motion – As time goes on, the apparent motion of the stars remains the same (at 15 degrees per hour westward).

PHYSICAL SETTING
EARTH SCIENCE — REFERENCE TABLE
2001 EDITION

Contents

PHYSICAL CONSTANTS

Radioactive Decay Data

RADIOACTIVE ISOTOPE	DISINTEGRATION	HALF-LIFE (years)
Carbon-14	$C^{14} \rightarrow N^{14}$	5.7×10^3
Potassium-40	$K^{40} \rightarrow Ar^{40}$ $\searrow Ca^{40}$	1.3×10^9
Uranium-238	$U^{238} \rightarrow Pb^{206}$	4.5×10^9
Rubidium-87	$Rb^{87} \rightarrow Sr^{87}$	4.9×10^{10}

Specific Heats of Common Materials

MATERIAL		SPECIFIC HEAT (calories/gram • C°)
Water	solid	0.5
	liquid	1.0
	gas	0.5
Dry air		0.24
Basalt		0.20
Granite		0.19
Iron		0.11
Copper		0.09
Lead		0.03

Properties of Water

Energy gained during melting	80 calories/gram
Energy released during freezing	80 calories/gram
Energy gained during vaporization	540 calories/gram
Energy released during condensation	540 calories/gram
Density at 3.98°C	1.00 gram/milliliter

EQUATIONS

Percent deviation from accepted value

$$\text{deviation (\%)} = \frac{\text{difference from accepted value}}{\text{accepted value}} \times 100$$

Eccentricity of an ellipse

$$\text{eccentricity} = \frac{\text{distance between foci}}{\text{length of major axis}}$$

Gradient

$$\text{gradient} = \frac{\text{change in field value}}{\text{distance}}$$

Rate of change

$$\text{rate of change} = \frac{\text{change in field value}}{\text{time}}$$

Density of a substance

$$\text{density} = \frac{\text{mass}}{\text{volume}}$$

Generalized Landscape Regions of New York State

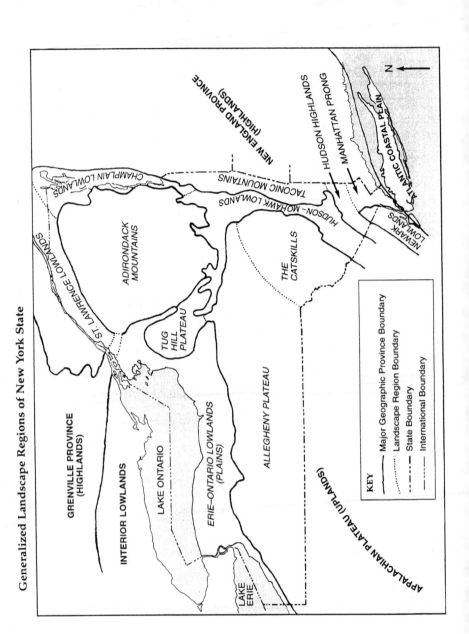

KEY

——— Major Geographic Province Boundary

········· Landscape Region Boundary

—··—··— State Boundary

------- International Boundary

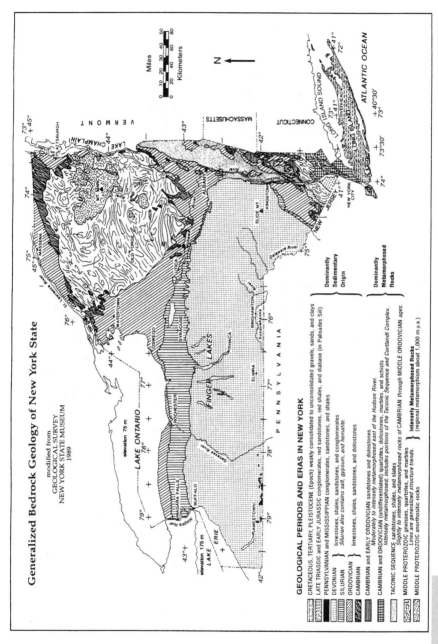

Generalized Bedrock Geology of New York State

modified from
GEOLOGICAL SURVEY
NEW YORK STATE MUSEUM
1989

GEOLOGICAL PERIODS AND ERAS IN NEW YORK

CRETACEOUS, TERTIARY, PLEISTOCENE (Epoch) weakly consolidated to unconsolidated gravels, sands, and clays

LATE TRIASSIC and EARLY JURASSIC conglomerates, red sandstones, red shales, and diabase (in Palisades Sill)

PENNSYLVANIAN and MISSISSIPPIAN conglomerates, sandstones, and shales

DEVONIAN limestones, shales, sandstones, and conglomerates

SILURIAN limestones, shales, sandstones, and conglomerates
  Silurian also contains salt, gypsum, and hematite.

CAMBRIAN limestones, shales, sandstones, and dolostones

CAMBRIAN and EARLY ORDOVICIAN sandstones and dolostones
  Moderately to intensely metamorphosed east of the Hudson River.

CAMBRIAN and ORDOVICIAN (undifferentiated) quartzites, dolostones, marbles, and schists
  Intensely metamorphosed, includes portions of the Taconic Sequence and Cortlandt Complex.

TACONIC SEQUENCE sandstones, quartzites, shales, and slates
  Slightly to intensely metamorphosed rocks of CAMBRIAN through MIDDLE ORDOVICIAN ages.

Intensely Metamorphosed Rocks of CAMBRIAN through MIDDLE ORDOVICIAN ages.

MIDDLE PROTEROZOIC gneisses, quartzites, and marbles

Intensely Metamorphosed Rocks
  (regional metamorphism about 1,000 m.y.a.)

MIDDLE PROTEROZOIC anorthositic rocks

Lines are generalized structure trends.

Dominantly Sedimentary Origin

Dominantly Metamorphosed Rocks

Surface Ocean Currents

WARM CURRENTS
COOL CURRENTS

Tectonic Plates

KEY:

Divergent Plate Boundary (usually broken by transform faults along mid-ocean ridges)

Transform Plate Boundary (Transform Fault)

Convergent Plate Boundary (Subduction Zone)

overriding plate
subducting plate

Complex or Uncertain Plate Boundary

Relative Motion at Plate Boundary

☆ Mantle Hot Spot

≁≁≁≁ Mid-Ocean Ridge

Rock Cycle in Earth's Crust

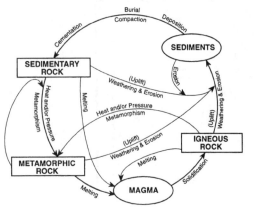

Relationship of Transported Particle Size to Water Velocity

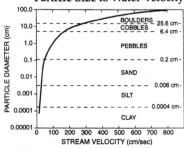

*This generalized graph shows the water velocity needed to maintain, but not start, movement. Variations occur due to differences in particle density and shape.

Scheme for Igneous Rock Identification

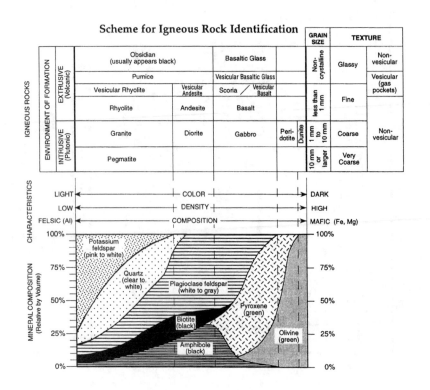

Scheme for Sedimentary Rock Identification

INORGANIC LAND-DERIVED SEDIMENTARY ROCKS					
TEXTURE	GRAIN SIZE	COMPOSITION	COMMENTS	ROCK NAME	MAP SYMBOL
Clastic (fragmental)	Pebbles, cobbles, and/or boulders embedded in sand, silt, and/or clay	Mostly quartz, feldspar, and clay minerals; may contain fragments of other rocks and minerals	Rounded fragments	Conglomerate	
			Angular fragments	Breccia	
	Sand (0.2 to 0.006 cm)		Fine to coarse	Sandstone	
	Silt (0.006 to 0.0004 cm)		Very fine grain	Siltstone	
	Clay (less than 0.0004 cm)		Compact; may split easily	Shale	

CHEMICALLY AND/OR ORGANICALLY FORMED SEDIMENTARY ROCKS					
TEXTURE	GRAIN SIZE	COMPOSITION	COMMENTS	ROCK NAME	MAP SYMBOL
Crystalline	Varied	Halite	Crystals from chemical precipitates and evaporites	Rock Salt	
	Varied	Gypsum		Rock Gypsum	
	Varied	Dolomite		Dolostone	
Bioclastic	Microscopic to coarse	Calcite	Cemented shell fragments or precipitates of biologic origin	Limestone	
	Varied	Carbon	From plant remains	Coal	

Scheme for Metamorphic Rock Identification

TEXTURE		GRAIN SIZE	COMPOSITION	TYPE OF METAMORPHISM	COMMENTS	ROCK NAME	MAP SYMBOL
FOLIATED	MINERAL ALIGNMENT	Fine	MICA QUARTZ FELDSPAR AMPHIBOLE GARNET PYROXENE	Regional (Heat and pressure increase with depth)	Low-grade metamorphism of shale	Slate	
		Fine to medium			Foliation surfaces shiny from microscopic mica crystals	Phyllite	
					Platy mica crystals visible from metamorphism of clay or feldspars	Schist	
	BANDING	Medium to coarse			High-grade metamorphism; some mica changed to feldspar; segregated by mineral type into bands	Gneiss	
NONFOLIATED		Fine	Variable	Contact (Heat)	Various rocks changed by heat from nearby magma/lava	Hornfels	
		Fine to coarse	Quartz	Regional or Contact	Metamorphism of quartz sandstone	Quartzite	
			Calcite and/or dolomite		Metamorphism of limestone or dolostone	Marble	
		Coarse	Various minerals in particles and matrix		Pebbles may be distorted or stretched	Metaconglomerate	

GEOLOGIC HISTORY OF NEW YORK STATE

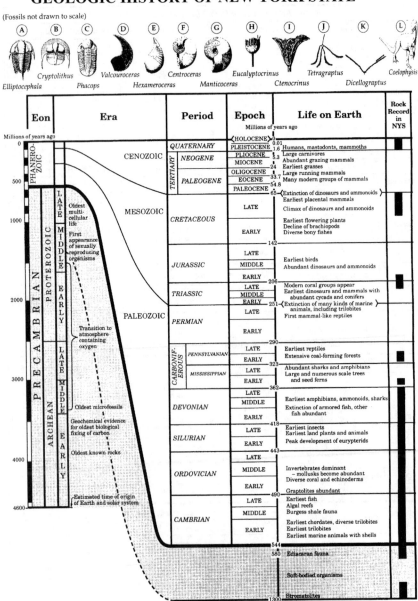

(Fossils not drawn to scale)

A · B · C · D · E · F · G · H · I · J · K · L

Elliptocephala — Cryptolithus — Valcouroceras — Centroceras — Eucalyptocrinus — Tetragraptus — Coelophysis
Phacops — Hexameroceras — Manticoceras — Ctenocrinus — Dicellograptus

Eon	Era		Period	Epoch	Life on Earth	Rock Record in NYS
				Millions of years ago		
				HOLOCENE 0		
	CENOZOIC	TERTIARY	QUATERNARY	PLEISTOCENE 0.01 / 1.6	Humans, mastodonts, mammoths	
			NEOGENE	PLIOCENE 5.3	Large carnivores	
				MIOCENE 24	Abundant grazing mammals / Earliest grasses	
			PALEOGENE	OLIGOCENE 33.7	Large running mammals	
				EOCENE 54.8	Many modern groups of mammals	
				PALEOCENE 65	Extinction of dinosaurs and ammonoids	
PHANEROZOIC	MESOZOIC		CRETACEOUS	LATE	Earliest placental mammals / Climax of dinosaurs and ammonoids	
				EARLY	Earliest flowering plants / Decline of brachiopods / Diverse bony fishes	
				142		
			JURASSIC	LATE	Earliest birds	
				MIDDLE	Abundant dinosaurs and ammonoids	
				EARLY 206		
			TRIASSIC	LATE	Modern coral groups appear / Earliest dinosaurs and mammals with / abundant cycads and conifers	
				MIDDLE		
				EARLY 251	Extinction of many kinds of marine animals, including trilobites	
	PALEOZOIC		PERMIAN	LATE	First mammal-like reptiles	
				EARLY 290		
			CARBONIFEROUS — PENNSYLVANIAN	LATE	Earliest reptiles	
				EARLY 323	Extensive coal-forming forests	
			CARBONIFEROUS — MISSISSIPPIAN	LATE	Abundant sharks and amphibians / Large and numerous scale trees / and seed ferns	
				EARLY 362		
			DEVONIAN	LATE		
				MIDDLE	Earliest amphibians, ammonoids, sharks / Extinction of armored fish, other / fish abundant	
				EARLY 418		
			SILURIAN	LATE	Earliest insects / Earliest land plants and animals	
				EARLY 443	Peak development of eurypterids	
			ORDOVICIAN	LATE	Earliest fish	
				MIDDLE	Invertebrates dominant / – mollusks become abundant / Diverse coral and echinoderms	
				EARLY 490	Graptolites abundant	
			CAMBRIAN	LATE	Earliest fish / Algal reefs	
				MIDDLE	Burgess shale fauna	
				EARLY	Earliest chordates, diverse trilobites / Earliest trilobites / Earliest marine animals with shells	
				544		
PROTEROZOIC				580	Ediacaran fauna	
					Soft-bodied organisms	
ARCHEAN				1300	Stromatolites	

Millions of years ago
0
500
1000
2000
3000
4000
4600

PRECAMBRIAN
PROTEROZOIC — LATE / MIDDLE / EARLY
ARCHEAN — LATE / MIDDLE / EARLY

Oldest multicellular life

First appearance of sexually reproducing organisms

Transition to atmosphere containing oxygen

Oldest microfossils

Geochemical evidence for oldest biological fixing of carbon

Oldest known rocks

Estimated time of origin of Earth and solar system

GEOLOGIC HISTORY OF NEW YORK STATE

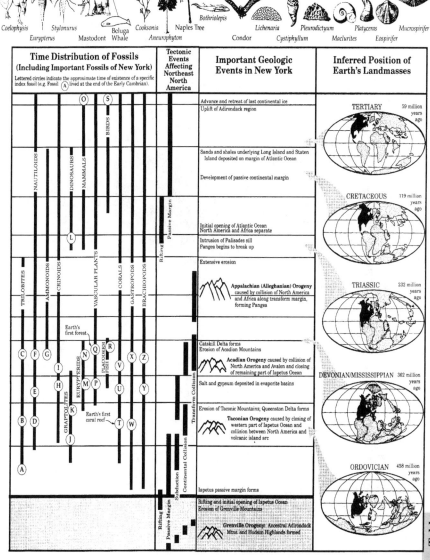

The Time Distribution of Fossils (Including Important Fossils of New York). Lettered circles indicate the approximate time of existence of a specific index fossil (e.g. Fossil Ⓐ lived at the end of the Early Cambrian).

Fossil index letters: L Coelophysis, M Eurypterus, N Stylonurus, O Mastodont / Beluga Whale, P Cooksonia, Q Aneurophyton / Naples Tree, R Bothriolepis, S Condor, T Lichenaria, U Cystiphyllum, V Pleurodictyum, W Maclurites, X Platyceras, Y Eospirifer, Z Mucrospirifer

Reference Table

Inferred Properties of Earth's Interior

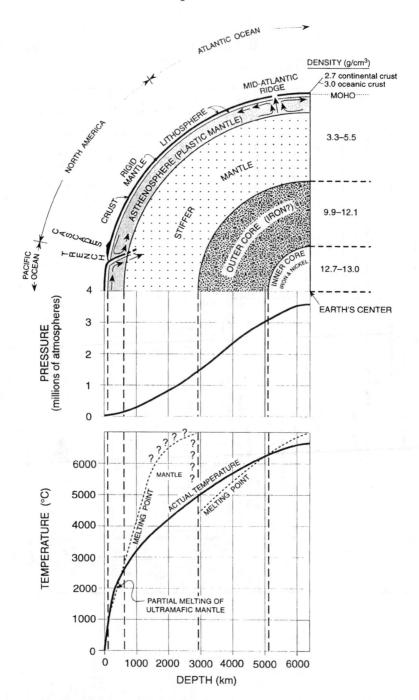

Average Chemical Composition
of Earth's Crust, Hydrosphere, and Troposphere

ELEMENT (symbol)	CRUST		HYDROSPHERE	TROPOSPHERE
	Percent by Mass	Percent by Volume	Percent by Volume	Percent by Volume
Oxygen (O)	46.40	94.04	33.0	21.0
Silicon (Si)	28.15	0.88		
Aluminum (Al)	8.23	0.48		
Iron (Fe)	5.63	0.49		
Calcium (Ca)	4.15	1.18		
Sodium (Na)	2.36	1.11		
Magnesium (Mg)	2.33	0.33		
Potassium (K)	2.09	1.42		
Nitrogen (N)				78.0
Hydrogen (H)			66.0	
Other	0.66	0.07	1.0	1.0

Earthquake P-wave and S-wave Travel Time

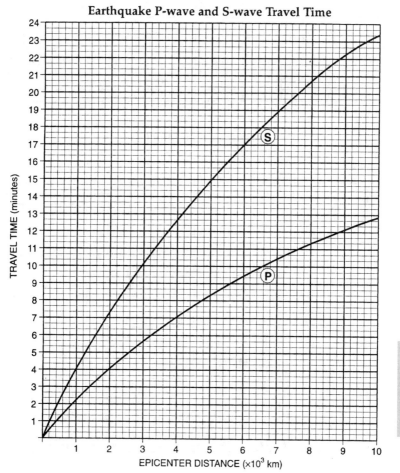

Reference Table

Dewpoint Temperatures (°C)

Dry-Bulb Temperature (°C)	Difference Between Wet-Bulb and Dry-Bulb Temperatures (C°)															
	0	1	2	3	4	5	6	7	8	9	10	11	12	13	14	15
−20	−20	−33														
−18	−18	−28														
−16	−16	−24														
−14	−14	−21	−36													
−12	−12	−18	−28													
−10	−10	−14	−22													
−8	−8	−12	−18	−29												
−6	−6	−10	−14	−22												
−4	−4	−7	−12	−17	−29											
−2	−2	−5	−8	−13	−20											
0	0	−3	−6	−9	−15	−24										
2	2	−1	−3	−6	−11	−17										
4	4	1	−1	−4	−7	−11	−19									
6	6	4	1	−1	−4	−7	−13	−21								
8	8	6	3	1	−2	−5	−9	−14								
10	10	8	6	4	1	−2	−5	−9	−14	−28						
12	12	10	8	6	4	1	−2	−5	−9	−16						
14	14	12	11	9	6	4	1	−2	−5	−10	−17					
16	16	14	13	11	9	7	4	1	−1	−6	−10	−17				
18	18	16	15	13	11	9	7	4	2	−2	−5	−10	−19			
20	20	19	17	15	14	12	10	7	4	2	−2	−5	−10	−19		
22	22	21	19	17	16	14	12	10	8	5	3	−1	−5	−10	−19	
24	24	23	21	20	18	16	14	12	10	8	6	2	−1	−5	−10	−18
26	26	25	23	22	20	18	17	15	13	11	9	6	3	0	−4	−9
28	28	27	25	24	22	21	19	17	16	14	11	9	7	4	1	−3
30	30	29	27	26	24	23	21	19	18	16	14	12	10	8	5	1

Relative Humidity (%)

Dry-Bulb Temperature (°C)	Difference Between Wet-Bulb and Dry-Bulb Temperatures (C°)															
	0	1	2	3	4	5	6	7	8	9	10	11	12	13	14	15
−20	100	28														
−18	100	40														
−16	100	48														
−14	100	55	11													
−12	100	61	23													
−10	100	66	33													
−8	100	71	41	13												
−6	100	73	48	20												
−4	100	77	54	32	11											
−2	100	79	58	37	20	1										
0	100	81	63	45	28	11										
2	100	83	67	51	36	20	6									
4	100	85	70	56	42	27	14									
6	100	86	72	59	46	35	22	10								
8	100	87	74	62	51	39	28	17	6							
10	100	88	76	65	54	43	33	24	13	4						
12	100	88	78	67	57	48	38	28	19	10	2					
14	100	89	79	69	60	50	41	33	25	16	8	1				
16	100	90	80	71	62	54	45	37	29	21	14	7	1			
18	100	91	81	72	64	56	48	40	33	26	19	12	6			
20	100	91	82	74	66	58	51	44	36	30	23	17	11	5		
22	100	92	83	75	68	60	53	46	40	33	27	21	15	10	4	
24	100	92	84	76	69	62	55	49	42	36	30	25	20	14	9	4
26	100	92	85	77	70	64	57	51	45	39	34	28	23	18	13	9
28	100	93	86	78	71	65	59	53	47	42	36	31	26	21	17	12
30	100	93	86	79	72	66	61	55	49	44	39	34	29	25	20	16

Temperature

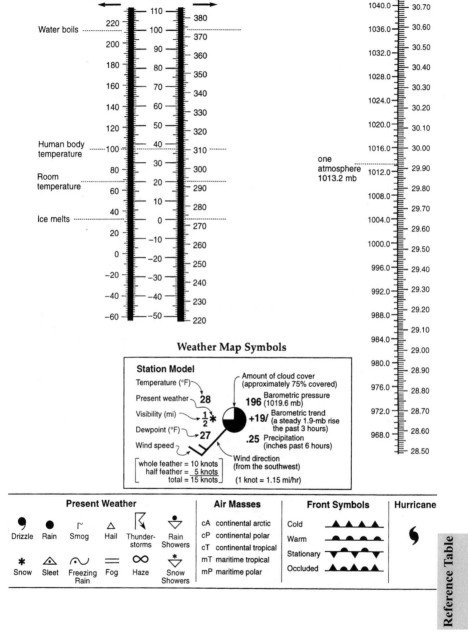

Fahrenheit | **Celsius** | **Kelvin**

Water boils — 220 / 200 / 100 / 370

Human body temperature — 100 / 40 / 310

Room temperature — 80 / 30 / 300 ... 20 / 290

Ice melts — 40 / 0 / 270

Pressure

millibars / inches

one atmosphere 1013.2 mb — 1012.0 / 29.90

Weather Map Symbols

Station Model

Temperature (°F) — **28**
Present weather
Visibility (mi) — ½ ✳
Dewpoint (°F) — **27**
Wind speed

Amount of cloud cover (approximately 75% covered)
196 Barometric pressure (1019.6 mb)
+19/ Barometric trend (a steady 1.9-mb rise the past 3 hours)
.25 Precipitation (inches past 6 hours)
Wind direction (from the southwest)
(1 knot = 1.15 mi/hr)

whole feather = 10 knots
half feather = 5 knots
total = 15 knots

Present Weather

Drizzle, Rain, Smog, Hail, Thunderstorms, Rain Showers
Snow, Sleet, Freezing Rain, Fog, Haze, Snow Showers

Air Masses

cA continental arctic
cP continental polar
cT continental tropical
mT maritime tropical
mP maritime polar

Front Symbols

Cold
Warm
Stationary
Occluded

Hurricane

Reference Table

Selected Properties of Earth's Atmosphere

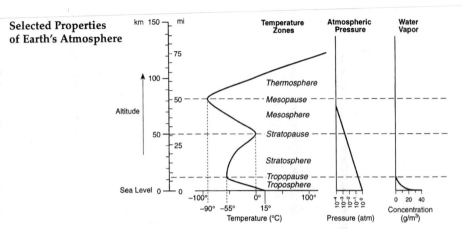

Electromagnetic Spectrum

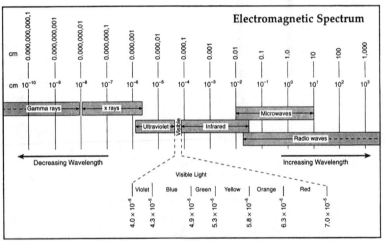

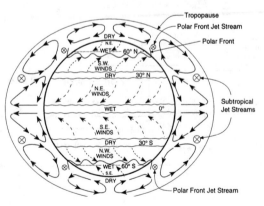

Planetary Wind and Moisture Belts in the Troposphere

The drawing to the left shows the locations of the belts near the time of an equinox. The locations shift somewhat with the changing latitude of the Sun's vertical ray. In the Northern Hemisphere, the belts shift northward in summer and southward in winter.

Luminosity and Temperature of Stars

(Name in italics refers to star shown by a ⊕)

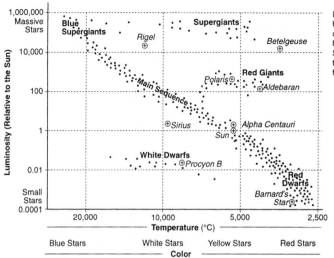

Luminosity is the brightness of stars compared to the brightness of our Sun as seen from the same distance from the observer.

Solar System Data

Object	Mean Distance from Sun (millions of km)	Period of Revolution	Period of Rotation	Eccentricity of Orbit	Equatorial Diameter (km)	Mass (Earth = 1)	Density (g/cm³)	Number of Moons
SUN	—	—	27 days	—	1,392,000	333,000.00	1.4	—
MERCURY	57.9	88 days	59 days	0.206	4,880	0.553	5.4	0
VENUS	108.2	224.7 days	243 days	0.007	12,104	0.815	5.2	0
EARTH	149.6	365.26 days	23 hr 56 min 4 sec	0.017	12,756	1.00	5.5	1
MARS	227.9	687 days	24 hr 37 min 23 sec	0.093	6,787	0.1074	3.9	2
JUPITER	778.3	11.86 years	9 hr 50 min 30 sec	0.048	142,800	317.896	1.3	16
SATURN	1,427	29.46 years	10 hr 14 min	0.056	120,000	95.185	0.7	18
URANUS	2,869	84.0 years	17 hr 14 min	0.047	51,800	14.537	1.2	21
NEPTUNE	4,496	164.8 years	16 hr	0.009	49,500	17.151	1.7	8
PLUTO	5,900	247.7 years	6 days 9 hr	0.250	2,300	0.0025	2.0	1
EARTH'S MOON	149.6 (0.386 from Earth)	27.3 days	27 days 8 hr	0.055	3,476	0.0123	3.3	—

Properties of Common Minerals

LUSTER	HARD-NESS	CLEAVAGE	FRACTURE	COMMON COLORS	DISTINGUISHING CHARACTERISTICS	USE(S)	MINERAL NAME	COMPOSITION*
Metallic Luster	1–2	✔		silver to gray	black streak, greasy feel	pencil lead, lubricants	**Graphite**	C
Metallic Luster	2.5	✔		metallic silver	very dense (7.6 g/cm³), gray-black streak	ore of lead	**Galena**	PbS
Metallic Luster	5.5–6.5		✔	black to silver	attracted by magnet, black streak	ore of iron	**Magnetite**	Fe_3O_4
Metallic Luster	6.5		✔	brassy yellow	green-black streak, cubic crystals	ore of sulfur	**Pyrite**	FeS_2
Either	1–6.5		✔	metallic silver or earthy red	red-brown streak	ore of iron	**Hematite**	Fe_2O_3
Nonmetallic Luster	1	✔		white to green	greasy feel	talcum powder, soapstone	**Talc**	$Mg_3Si_4O_{10}(OH)_2$
Nonmetallic Luster	2		✔	yellow to amber	easily melted, may smell	vulcanize rubber, sulfuric acid	**Sulfur**	S
Nonmetallic Luster	2	✔		white to pink or gray	easily scratched by fingernail	plaster of paris and drywall	**Gypsum** (Selenite)	$CaSO_4 \cdot 2H_2O$
Nonmetallic Luster	2–2.5	✔		colorless to yellow	flexible in thin sheets	electrical insulator	**Muscovite Mica**	$KAl_3Si_3O_{10}(OH)_2$
Nonmetallic Luster	2.5	✔		colorless to white	cubic cleavage, salty taste	food additive, melts ice	**Halite**	NaCl
Nonmetallic Luster	2.5–3	✔		black to dark brown	flexible in thin sheets	electrical insulator	**Biotite Mica**	$K(Mg,Fe)_3$ $AlSi_3O_{10}(OH)_2$
Nonmetallic Luster	3	✔		colorless or variable	bubbles with acid	cement, polarizing prisms	**Calcite**	$CaCO_3$
Nonmetallic Luster	3.5	✔		colorless or variable	bubbles with acid when powdered	source of magnesium	**Dolomite**	$CaMg(CO_3)_2$
Nonmetallic Luster	4	✔		colorless or variable	cleaves in 4 directions	hydrofluoric acid	**Fluorite**	CaF_2
Nonmetallic Luster	5–6	✔		black to dark green	cleaves in 2 directions at 90°	mineral collections	**Pyroxene** (commonly Augite)	$(Ca,Na) (Mg,Fe,Al)$ $(Si,Al)_2O_6$
Nonmetallic Luster	5.5	✔		black to dark green	cleaves at 56° and 124°	mineral collections	**Amphiboles** (commonly Hornblende)	$CaNa(Mg,Fe)_4 (Al,Fe,Ti)_3$ $Si_6O_{22}(O,OH)_2$
Nonmetallic Luster	6	✔		white to pink	cleaves in 2 directions at 90°	ceramics and glass	**Potassium Feldspar** (Orthoclase)	$KAlSi_3O_8$
Nonmetallic Luster	6	✔		white to gray	cleaves in 2 directions, striations visible	ceramics and glass	**Plagioclase Feldspar** (Na-Ca Feldspar)	$(Na,Ca)AlSi_3O_8$
Nonmetallic Luster	6.5		✔	green to gray or brown	commonly light green and granular	furnace bricks and jewelry	**Olivine**	$(Fe,Mg)_2SiO_4$
Nonmetallic Luster	7		✔	colorless or variable	glassy luster, may form hexagonal crystals	glass, jewelry, and electronics	**Quartz**	SiO_2
Nonmetallic Luster	7		✔	dark red to green	glassy luster, often seen as red grains in NYS metamorphic rocks	jewelry and abrasives	**Garnet** (commonly Almandine)	$Fe_3Al_2Si_3O_{12}$

*Chemical Symbols:

Al = aluminum	Cl = chlorine	H = hydrogen	Na = sodium	S = sulfur
C = carbon	F = fluorine	K = potassium	O = oxygen	Si = silicon
Ca = calcium	Fe = iron	Mg = magnesium	Pb = lead	Ti = titanium

✔ = dominant form of breakage

Metric Ruler

cm 1 2 3 4 5 6 7 8 9 10 11 12 13 14 15 16 17 18 19 20 21

Correlation of Questions to Topic Area

Astronomy
June 2005 – 1, 3, 4, 6, 9, 11, 48, 49, 50, 66, 67, 77, 78, 79
June 2006 – 1, 2, 5, 6, 7, 8, 10, 11, 39, 40, 41, 56, 57, 58, 59, 60, 71
June 2007 – 1, 2, 25, 27, 36, 37, 38, 39, 40, 41, 42, 45, 65, 66, 67, 68, 82
June 2008 – 1, 2, 3, 4, 5, 10, 11, 12, 13, 36, 37, 38, 39, 59, 60

Climate
June 2005 – 7, 8, 25, 68, 71
June 2006 – 17, 19, 66, 67, 69
June 2007 – 3, 4, 5, 28, 46, 47
June 2008 – 20, 27, 53, 64, 66, 67, 68

Dynamic Earth
June 2005 – 25, 26, 28, 55, 56, 57, 80, 81, 82, 83
June 2006 – 22, 24, 25, 32, 81, 82, 83
June 2007 – 19, 33, 75, 76, 78, 79, 80
June 2008 – 23, 24, 25, 61, 62, 68, 73

Energy – Insolation
June 2005 – 5, 10, 12, 27, 29, 74
June 2006 – 9, 14, 18, 47, 48, 49
June 2007 – 4, 6, 26, 37, 38, 82
June 2008 – 9, 19, 34, 41, 42, 43, 73, 75

Fields – Latitude/Longitude
June 2005 – 4, 38, 42, 43, 44
June 2006 – 2, 20, 30, 51, 52, 70, 71, 72
June 2007 – 70, 71, 72, 73, 77
June 2008 – 78, 79, 80

Geologic History
June 2005 – 31, 32, 33, 40, 41, 47, 58, 59, 60
June 2006 – 4, 23, 27, 38, 42, 43, 61, 77, 78, 79
June 2007 – 20, 21, 32, 35, 43, 44, 54, 55, 56, 57
June 2008 – 20, 21, 22, 45, 72

Landscape
June 2005 – 13, 34, 45, 70
June 2006 – 29, 35
June 2007 – 22, 61
June 2008 – 27, 28, 40, 48, 49

Maps – Charts
June 2005 – 16, 26, 27, 28, 30, 32, 33, 46, 47, 54, 66, 72
June 2006 – 3, 4, 13, 15, 25, 26, 36, 51, 63, 68
June 2007 – 8, 12, 13, 23, 33, 34, 39, 41, 42, 58
June 2008 – 6, 18, 21, 23, 24, 26, 33, 40, 42, 49, 62, 63, 66, 67, 68, 72

Rocks and Minerals
June 2005 – 19, 20, 21, 22, 23, 24, 30, 39, 75
June 2006 – 21, 28, 34, 37, 38, 53, 54, 55, 80
June 2007 – 14, 16, 18, 23, 24, 30, 31, 32, 62, 63, 64
June 2008 – 34, 35, 46, 47, 52, 69, 70, 71, 72

Seasons – Sun's Path
June 2005 – 2, 10, 48, 76, 77, 78, 79
June 2006 – 9
June 2007 – 27, 28, 37, 38, 82
June 2008 – 41, 74, 76, 77

Water
June 2005 – 17, 46
June 2006 – 15, 44, 45
June 2007 – 7, 8, 9, 54, 58
June 2008 – 9, 15, 18, 30, 33, 44, 50, 51, 53, 63, 64, 65

Weather
June 2005 – 14, 15, 35, 36, 37, 38, 51, 52, 53, 64, 65
June 2006 – 12, 13, 31, 33, 65, 66, 73, 74, 75, 76
June 2007 – 8, 10, 11, 13, 29, 51, 52, 53
June 2008 – 7, 8, 14, 15, 16, 17, 54, 55, 56, 57, 58

Weathering – Erosion – Deposition
June 2005 – 18, 61, 62, 63, 69, 73
June 2006 – 3, 16, 45, 46, 50, 62, 64
June 2007 – 15, 17, 22, 48, 49, 50, 59, 61
June 2008 – 29, 30, 31, 32, 33, 38, 48, 49, 50, 64